SCIENTIFIC AMERICAN EXPLORES BIG IDEAS

Is There More Than One Universe?

The Editors of *Scientific American*

SCIENTIFIC AMERICAN EDUCATIONAL PUBLISHING

New York

Published in 2024 by Scientific American Educational Publishing
in association with **The Rosen Publishing Group**
2544 Clinton Street, Buffalo NY 14224

Contains material from Scientific American®, a division of Springer Nature America, Inc., reprinted by permission, as well as original material from The Rosen Publishing Group®.

Copyright © 2024 Scientific American® and Rosen Publishing Group®.

All rights reserved.

First Edition

Scientific American
Lisa Pallatroni: Project Editor

Rosen Publishing
Daniel R. Faust: Compiling Editor
Michael Moy: Senior Graphic Designer

Cataloging-in-Publication Data

Names: Scientific American, Inc.
Title: Is there more than one universe? / edited by the Scientific American Editors.
Description: First Edition. | New York : Scientific American Educational Publishing, 2024. | Series: Scientific American explores big ideas | Includes glossary and index.
Identifiers: ISBN 9781725349728 (pbk.) | ISBN 9781725349735 (library bound)| ISBN 9781725349742 (ebook)
Subjects: LCSH: Multiverse–Juvenile literature. | Cosmology–Juvenile literature. | Quantum theory–Juvenile literature.
Classification: LCC QB983.I884 2024 | DDC 523.1–dc23

Manufactured in the United States of America
Websites listed were live at the time of publication.

Cover: NASA images/Shutterstock.com

CPSIA Compliance Information: Batch # SACS24.
For Further Information contact Rosen Publishing at 1-800-237-9932.

CONTENTS

Introduction 5

Section 1: Multiple Dimensions 6

1.1 The Case for Parallel Universes 7
By Alexander Vilenkin and Max Tegmark

1.2 The Great Cosmic Roller-Coaster Ride 16
By Cliff Burgess and Fernando Quevedo

1.3 Long Live the Multiverse! 26
By Tom Siegfried

1.4 Our Improbable Existence Is No Evidence for a Multiverse 30
By Philip Goff

1.5 Extra Dimensions 34
By George Musser

1.6 New Phase of Matter Opens Portal to Extra Time Dimension 37
By Zeeya Merali

Section 2: Time and Space 42

2.1 How Scientists Solved One of the Greatest Open Questions in Quantum Physics 43
By Spyridon Michalakis

2.2 Quantum Gravity in Flatland 59
By Steven Carlip

2.3 The Beauty at the Heart of a "Spooky" Mystery 70
By John Horgan

2.4 This Twist on Schrödinger's Cat Paradox Has Major Implications for Quantum Theory 74
By Zeeya Merali

2.5 Can Quantum Mechanics Save the Cosmic Multiverse? 84
By Yasunori Nomura

2.6 The Difficult Birth of the "Many Worlds" Interpretation of Quantum Mechanics 94
By Adam Becker

2.7 Black Hole Discovery Helps to Explain Quantum Nature of the Cosmos 98
By Edgar Shaghoulian

2.8 How the Inside of a Black Hole Is Secretly on the Outside 107
By Ahmed Almheiri

Section 3: Pulling Strings — 117

3.1 Will String Theory Finally Be Put to the Experimental Test? — 118
By Brendan Z. Foster

3.2 String Theory May Create Far Fewer Universes Than Thought — 122
By Clara Moskowitz and Lee Billings

Section 4: Searching for Answers — 127

4.1 Looking for Life in the Multiverse — 128
By Alejandro Jenkins, Gilad Perez

4.2 Does a Multiverse Fermi Paradox Disprove the Multiverse? — 139
By Caleb A. Scharf

4.3 Was Our Universe Created in a Laboratory? — 142
By Avi Loeb

4.4 Multiverse Theories Are Bad for Science — 145
By John Horgan

4.5 Life Quest: Could Parallel Universes Be Congenial to Life? — 149
By Mariette DiChristina

4.6 Multiverse Controversy Heats Up over Gravitational Waves — 151
By Clara Moskowitz

Glossary — 155
Further Information — 157
Citations — 158
Index — 159

INTRODUCTION

From movies like *Everything Everywhere All at Once* and *Spider-Man: No Way Home* to television series like *Loki* and *Rick and Morty*, it seems like the multiverse is everywhere. But, while it might feel like the concept of multiple parallel worlds is just the latest Hollywood fad, it's actually based on real scientific theory. As the name implies, the multiverse is a theoretical group of multiple universes that contain everything in existence. The earliest concept of a universe composed of infinite worlds dates back to the philosophers of ancient Greece. The modern notion of the multiverse didn't develop until the 20th century. The articles that follow take a detailed look at the multiverse and explore the science behind the theory.

In Section 1, "Multiple Dimensions," the basics of the multiverse are explored, including any scientific evidence that may prove or disprove its existence. Section 2, "Time and Space," serves as an introduction to quantum mechanics, one of the two concepts that form the core of modern physics. Section 3, "Pulling Strings," examines string theory, one of the concepts in theoretical physics that forms the foundation of the theory of the multiverse. Section 4, "Searching for Answers," raises a number of questions about the origins of the multiverse and the possibility of life in the other dimensions that could make up the multiverse.

Section 1: Multiple Dimensions

1.1 The Case for Parallel Universes
   By Alexander Vilenkin and Max Tegmark

1.2 The Great Cosmic Roller-Coaster Ride
   By Cliff Burgess and Fernando Quevedo

1.3 Long Live the Multiverse!
   By Tom Siegfried

1.4 Our Improbable Existence Is No Evidence for a Multiverse
   By Philip Goff

1.5 Extra Dimensions
   By George Musser

1.6 New Phase of Matter Opens Portal to Extra Time Dimension
   By Zeeya Merali

The Case for Parallel Universes

By Alexander Vilenkin and Max Tegmark

Welcome to the Multiverse

By Alexander Vilenkin

The universe as we know it originated in a great explosion that we call the big bang. For nearly a century cosmologists have been studying the aftermath of this explosion: how the universe expanded and cooled down, and how galaxies were gradually pulled together by gravity. The nature of the bang itself has come into focus only relatively recently. It is the subject of the theory of inflation, which was developed in the early 1980s by Alan Guth, Andrei Linde and others, and has led to a radically new global view of the universe.

Inflation is a period of super-fast, accelerated expansion in early cosmic history. It is so fast that in a fraction of a second a tiny subatomic speck of space is blown to dimensions much greater than the entire currently observable region. At the end of inflation, the energy that drove the expansion ignites a hot fireball of particles and radiation. This is what we call the big bang.

The end of inflation is triggered by quantum, probabilistic processes and does not occur everywhere at once. In our cosmic neighborhood, inflation ended 13.7 billion years ago, but it still continues in remote parts of the universe, and other "normal" regions like ours are constantly being formed. The new regions appear as tiny, microscopic bubbles and immediately start to grow. The bubbles keep growing without bound; in the meantime they are driven apart by the inflationary expansion, making room for more bubbles to form. This never-ending process is called eternal inflation. We live in one of the bubbles and can observe only a small part of it. No matter how fast we travel, we cannot catch up with the expanding boundaries of our bubble, so for all practical purposes we live in a self-contained bubble universe.

Is There More Than One Universe?

The theory of inflation explained some otherwise mysterious features of the big bang, which simply had to be postulated before. It also made a number of testable predictions, which were then spectacularly confirmed by observations. By now inflation has become the leading cosmological paradigm.

Another key aspect of the new worldview derives from string theory, which is at present our best candidate for the fundamental theory of nature. String theory admits an immense number of solutions describing bubble universes with diverse physical properties. The quantities we call constants of nature, such as the masses of elementary particles, Newton's gravitational constant, and so on, take different values in different bubble types. Now combine this with the theory of inflation. Each bubble type has a certain probability to form in the inflating space. So inevitably, an unlimited number of bubbles of all possible types will be formed in the course of eternal inflation.

This picture of the universe, or *multiverse*, as it is called, explains the long-standing mystery of why the constants of nature appear to be fine-tuned for the emergence of life. The reason is that intelligent observers exist only in those rare bubbles in which, by pure chance, the constants happen to be just right for life to evolve. The rest of the multiverse remains barren, but no one is there to complain about that.

Some of my physicist colleagues find the multiverse theory alarming. Any theory in physics stands or falls depending on whether its predictions agree with the data. But how can we verify the existence of other bubble universes? Paul Steinhardt and George Ellis have argued, for example, that the multiverse theory is unscientific, because it cannot be tested, even in principle.

Surprisingly, observational tests of the multiverse picture may in fact be possible. Anthony Aguirre, Matt Johnson, Matt Kleban and others have pointed out that a collision of our expanding bubble with another bubble in the multiverse would produce an imprint in the cosmic background radiation—a round spot of higher or lower radiation intensity. A detection of such a spot with the predicted intensity profile would provide direct evidence for the existence of other bubble universes. The search is now on, but unfortunately there is no guarantee that a bubble collision has occurred within our cosmic horizon.

There is also another approach that one can follow. The idea is to use our theoretical model of the multiverse to predict the constants of nature that we can expect to measure in our local region. If the constants vary from one bubble universe to another, their local values cannot be predicted with certainty, but we can still make *statistical* predictions. We can derive from the theory what values of the constants are most likely to be measured by a typical observer in the multiverse. Assuming that we are typical—the assumption that I called the *principle of mediocrity*—we can then predict the likely values of the constants in our bubble.

This strategy has been applied to the energy density of the vacuum, also known as "dark energy." Steven Weinberg has noted that in regions where dark energy is large, it causes the universe to expand very fast, preventing matter from clumping into galaxies and stars. Observers are not likely to evolve in such regions. Calculations showed that most galaxies (and therefore most observers) are in regions where the dark energy is about the same as the density of matter at the epoch of galaxy formation. The prediction is therefore that a similar value should be observed in our part of the universe.

For the most part, physicists did not take these ideas seriously, but much to their surprise, dark energy of roughly the expected magnitude was detected in astronomical observations in the late 1990s. This could be our first evidence that there is indeed a huge multiverse out there. It has changed many minds.

The multiverse theory is still in its infancy, and some conceptual problems remain to be resolved. But, as Leonard Susskind wrote, "I would bet that at the turn of the 22nd century philosophers and physicists will look nostalgically at the present and recall a golden age in which the narrow provincial 20th century concept of the universe gave way to a bigger better [multiverse] ... of mind-boggling proportions."

Is There More Than One Universe?

The Multiverse Strikes Back

By Max Tegmark

Inspired by an interesting critique of multiverses in the August issue of *Scientific American*, penned by relativity pioneer George F. R. Ellis, let my give you my two cents' worth.

Multiverse ideas have traditionally received short shrift from the establishment: Giordano Bruno with his infinite-space multiverse got burned at the stake in 1600 and Hugh Everett with his quantum multiverse got burned on the physics job market in 1957. I've even felt some of the heat first-hand, with senior colleagues suggesting that my multiverse-related publications were nuts and would ruin my career. There's been a sea-change in recent years, however. Parallel universes are now all the rage, cropping up in books, movies and even jokes: "You passed your exam in many parallel universes—but not this one."

This airing of ideas certainly hasn't led to a consensus among scientists, but it's made the multiverse debate much more nuanced and, in my opinion, more interesting, with scientists moving beyond shouting sound bites past each other and genuinely trying to understand opposing points of view. George Ellis's new article is a great example of this, and I highly recommend reading it if you haven't already.

By our universe, I mean the spherical region of space from which light has had time to reach us during the 13.7 billion years since our big bang. When talking about parallel universes, I find it useful to distinguish between four different levels: Level I (other such regions far away in space where the apparent laws of physics are the same, but where history played out differently because things started out differently), Level II (regions of space where even the apparent laws of physics are different), Level III (parallel worlds elsewhere in the so-called Hilbert space where quantum reality plays out), and Level IV (totally disconnected realities governed by different mathematical equations).

Section 1: Multiple Dimensions

In his critique, George classifies many of the arguments in favor of these multiverse levels and argues that they all have problems. Here's my summary of his main anti-multiverse arguments:

1. Inflation may be wrong (or not eternal)
2. Quantum mechanics may be wrong (or not unitary)
3. String theory may be wrong (or lack multiple solutions)
4. Multiverses may be unfalsifiable
5. Some claimed multiverse evidence is dubious
6. Fine-tuning arguments may assume too much
7. It's a slippery slope to even bigger multiverses

(George didn't actually mention (2) in the article, but I'm adding it here because I think he would have if the editor had allowed him more than six pages.)

What's my take on this critique? Interestingly, I agree with all of these seven statements—and nonetheless, I'll still happily bet my life savings on the existence of a multiverse!

Let's start with the first four. Inflation naturally produces the Level I multiverse, and if you add in string theory with a landscape of possible solutions, you get Level II, too. Quantum mechanics in its mathematically simplest ("unitary") form gives you Level III. So if these theories are ruled out, then key evidence for these multiverses collapses.

Remember: Parallel universes are not a theory—they are predictions of certain theories.

To me, the key point is that if theories are scientific, then it's legitimate science to work out and discuss all their consequences even if they involve unobservable entities. For a theory to be falsifiable, we need not be able to observe and test all its predictions, merely at least one of them. My answer to (4) is therefore that what's scientifically testable are our mathematical theories, not necessarily their implications, and that this is quite OK. For example, because Einstein's theory of general relativity has successfully predicted many things that we can observe, we also take seriously its predictions for things we cannot observe, e.g., what happens inside black holes.

Likewise, if we're impressed by the successful predictions of inflation or quantum mechanics so far, then we need to take seriously also their other predictions, including the Level I and Level III multiverse. George even mentions the possibility that eternal inflation may one day be ruled out—to me, this is simply an argument that eternal inflation is a scientific theory.

String theory certainly hasn't come as far as inflation and quantum mechanics in terms of establishing itself as a testable scientific theory. However, I suspect that we'll be stuck with a Level II multiverse even if string theory turns out to be a red herring. It's quite common for mathematical equations to have multiple solutions, and as long as the fundamental equations describing our reality do, then eternal inflation generically creates huge regions of space that physically realize each of these solutions. For example, the equations governing water molecules, which have nothing to do with string theory, permit the three solutions corresponding to steam, liquid water and ice, and if space itself can similarly exist in different phases, inflation will tend to realize them all.

George lists a number of observations purportedly supporting multiverse theories that are dubious at best, like evidence that certain constants of nature aren't really constant, evidence in the cosmic microwave background radiation of collisions with other universes or strangely connected space, etc. I totally share his skepticism to these claims. In all these cases, however, the controversies have been about the analysis of the data, much like in the cold fusion debacle. To me, the very fact that scientists are making these measurements and arguing about data details is further evidence that this is within the pale of science: this is precisely what separates a scientific controversy from a nonscientific one!

Our universe appears surprisingly fine-tuned for life in the sense that if you tweaked many of our constants of nature by just a tiny amount, life as we know it would be impossible. Why? If there's a Level II multiverse where these "constants" take all

Section 1: Multiple Dimensions

possible values, it's not surprising that we find ourselves in one of the rare universes that are inhabitable, just like it's not surprising that we find ourselves living on Earth rather than Mercury or Neptune. George objects to the fact that you need to assume a multiverse theory to draw this conclusion, but that's how we test any scientific theory: we assume that it's true, work out the consequences, and discard the theory if the predictions fail to match the observations. Some of the fine-tuning appears extreme enough to be quite embarrassing—for example, we need to tune the dark energy to about 123 decimal places to make habitable galaxies. To me, an unexplained coincidence can be a tell-tale sign of a gap in our scientific understanding. Dismissing it by saying "We just got lucky—now stop looking for an explanation!" is not only unsatisfactory, but is also tantamount to ignoring a potentially crucial clue.

George argues that if we take seriously that anything that could happen does happen, we're led down a slippery slope toward even larger multiverses, like the Level IV one. Since this is my favorite multiverse level, and I'm one of the very few proponents of it, this is a slope that I'm happy to slide down!

George also mentions that multiverses may fall foul of Occam's razor by introducing unnecessary complications. As a theoretical physicist, I judge the elegance and simplicity of a theory not by its ontology, but by the elegance and simplicity of its mathematical equations—and it's quite striking to me that the mathematically simplest theories tend to give us multiverses. It's proven remarkably hard to write down a theory which produces exactly the universe we see and nothing more.

Finally, there's an anti-multiverse argument which I commend George for avoiding, but which is in my opinion the most persuasive one of all for most people: the parallel universes just seems too weird to be true.

Having looked at anti-multiverse arguments, let's now analyze the pro-multiverse case a bit more closely. I'm going to argue that all the controversial issues melt away if we accept the External Reality

Is There More Than One Universe?

Hypothesis: there exists an external physical reality completely independent of us humans. Suppose that this hypothesis is correct. Then most multiverse critique rests on some combination of the following three dubious assumptions:

1. Omnivision assumption: physical reality must be such that at least one observer can in principle observe all of it.
2. Pedagogical reality assumption: physical reality must be such that all reasonably informed human observers feel they intuitively understand it.
3. No-copy assumption: no physical process can copy observers or create subjectively indistinguishable observers.

(1) and (2) appear to be motivated by little more than human hubris. The omnivision assumption effectively redefines the word "exists" to be synonymous with what is observable to us humans, akin to an ostrich with its head in the sand. Those who insist on the pedagogical reality assumption will typically have rejected comfortingly familiar childhood notions like Santa Claus, local realism, the Tooth Fairy, and creationism—but have they really worked hard enough to free themselves from comfortingly familiar notions that are more deeply rooted? In my personal opinion, our job as scientists is to try to figure out how the world works, not to tell it how to work based on our philosophical preconceptions.

If the omnivision assumption is false, then there are unobservable things that exist and we live in a multiverse.

If the pedagogical reality assumption is false, then the objection that multiverses are too weird makes no logical sense.

If the no-copy assumption is false, then there's no fundamental reason why there can't be copies of you elsewhere in the external reality—indeed, both eternal inflation and unitary quantum mechanics provide mechanisms for creating them.

We humans have a well-documented tendency toward hubris, arrogantly imagining ourselves at center stage, with everything revolving around us. We've gradually learned that it's instead we who are revolving around the sun, which is itself revolving around one galaxy among countless others. Thanks to breakthroughs in physics, we may be gaining still deeper insights into the very nature of reality.

The price we have to pay is becoming more humble—which will probably do us good—but in return we may find ourselves inhabiting a reality grander than our ancestors dreamed of in their wildest dreams.

Referenced

Many Worlds in One: The Search for Other Universes. Alex Vilenkin. Hill and Wang, 2006.

The Cosmic Landscape: String Theory and the Illusion of Intelligent Design. Leonard Susskind. Back Bay Books, 2006.

The Hidden Reality: Parallel Universes and the Hidden Laws of the Cosmos. Brian Greene. Knopf, 2011.

About the Author

Known as "Mad Max" for his unorthodox ideas and passion for adventure, Max Tegmark's scientific interests range from precision cosmology to the ultimate nature of reality, all explored in his new popular book, "Our Mathematical Universe." He is an MIT physics professor with more than 200 technical papers credit, and he has been featured in dozens of science documentaries. His work with the SDSS collaboration on galaxy clustering shared the first prize in Science *magazine's "Breakthrough of the Year: 2003."*

The Great Cosmic Roller-Coaster Ride

By Cliff Burgess and Fernando Quevedo

You might not think that cosmologists could feel claustrophobic in a universe that is 46 billion light-years in radius and filled with sextillions of stars. But one of the emerging themes of 21st-century cosmology is that the known universe, the sum of all we can see, may just be a tiny region in the full extent of space. Various types of parallel universes that make up a grand "multiverse" often arise as side effects of cosmological theories. We have little hope of ever directly observing those other universes, though, because they are either too far away or somehow detached from our own universe.

Some parallel universes, however, could be separate from but still able to interact with ours, in which case we could detect their direct effects. The possibility of these worlds came to cosmologists' attention by way of string theory, the leading candidate for the foundational laws of nature. Although the eponymous strings of string theory are extremely small, the principles governing their properties also predict new kinds of larger membranelike objects—"branes," for short. In particular, our universe may be a three-dimensional brane in its own right, living inside a nine-dimensional space. The reshaping of higher-dimensional space and collisions between different universes may have led to some of the features that astronomers observe today.

String theory has received some unfavorable press of late. The criticisms are varied and beyond the scope of this article, but the most pertinent is that it has yet to be tested experimentally. That is a legitimate worry. It is less a criticism of string theory, though, than a restatement of the general difficulty of testing theories about extremely small scales. All proposed foundational laws encounter the same problem, including other proposals such as loop quantum gravity. String theorists continue to seek ways to test their theory. One approach with promise is to study how it might explain mysterious aspects of our universe, foremost among which is the way the pace of cosmic expansion has changed over time.

Going for a Ride

Next year will be the 10th anniversary of the announcement that the universe is expanding at an ever quickening pace, driven by some unidentified constituent known as dark energy. Most cosmologists think that an even faster period of accelerated expansion, known as inflation, also occurred long before atoms, let alone galaxies, came into being. The universe's temperature shortly after this early inflationary period was billions of times higher than any yet observed on Earth. Cosmologists and elementary particle physicists find themselves making common cause to try to learn the fundamental laws of physics at such high temperatures. This cross-fertilization of ideas is stimulating a thorough rethinking of the early universe in terms of string theory.

The concept of inflation emerged to explain a number of simple yet puzzling observations. Many of these involve the cosmic microwave background radiation (CMBR), a fossil relic of the hot early universe. For instance, the CMBR reveals that our early universe was almost perfectly uniform—which is strange because none of the usual processes that homogenize matter (such as fluid flow) would have had time to operate. In the early 1980s Alan H. Guth, now at the Massachusetts Institute of Technology, found that an extremely rapid period of expansion could account for this homogeneity. Such an accelerating expansion diluted any preexisting matter and smoothed out deviations in density.

Equally important, it did not make the universe exactly homogeneous. The energy density of space during the inflationary period fluctuated because of the intrinsically statistical quantum laws that govern nature over subatomic distances. Like a giant photocopy machine, inflation enlarged these small quantum fluctuations to astronomical size, giving rise to predictable fluctuations in density later in cosmic history.

What is seen in the CMBR reproduces the predictions of inflationary theory with spectacular accuracy. This observational success has made inflation the leading proposal for how the universe behaved at very early times. Upcoming satellites, such as the European Space Agency's Planck observatory that is scheduled for launch next year, will look for corroborating evidence.

But do the laws of physics actually produce this inflation? Here the story gets murkier. It is notoriously difficult to get a universe full of regular forms of matter to accelerate in its expansion. Such a speedup takes a type of energy with a very unusual set of properties: its energy density must be positive and remain almost constant even as the universe dramatically expands, but the energy density must then suddenly decrease to allow inflation to end.

At first sight, it seems impossible for the energy density of anything to remain constant, because the expansion of space should dilute it. But a special source of energy, called a scalar field, can avoid this dilution. You can think of a scalar field as an extremely primitive substance that fills space, rather like a gas, except that it does not behave like any gas you have ever seen. It is similar to but simpler than the better-known electromagnetic and gravitational fields. The term "scalar field" simply means that it is described by a single number, its magnitude, that can vary from location to location within space. In contrast, a magnetic field is a vector field, which has both a magnitude and a direction (toward the north magnetic pole) at each point in space. A weather report provides examples of both types of field: temperature and pressure are scalars, whereas wind velocity is a vector.

The scalar field that drove inflation, dubbed the "inflaton" field, evidently caused the expansion to accelerate for a long period before switching off abruptly. The dynamics were like the first moments of a roller-coaster ride. The coaster initially climbs slowly along a gentle hill. ("Slowly" is a relative term; the process was still very fast in human terms.) Then comes the breathtaking plunge during which potential energy is converted to kinetic energy and ultimately heat. This behavior is not easy to reproduce theoretically. Physicists have

Section 1: Multiple Dimensions

made a variety of proposals over the past 25 years, but none has yet emerged as compelling. The search is hampered by our ignorance of what might be going on at the incredibly high energies that are likely to be relevant.

Brane Bogglers

During the 1980s, as inflation was gaining credence, an independent line of reasoning was making progress toward reducing our ignorance on that very issue. String theory proposes that subatomic particles are actually tiny one-dimensional objects like miniature rubber bands. Some of these strings form loops (so-called closed strings), but others are short segments with two ends (open strings). The theory attributes all the elementary particles ever discovered, and many more undiscovered, to different styles of vibration of these types of strings. The best part of string theory is that, unlike other theories of elementary particles, it organically contains gravity within itself. That is, gravity emerges naturally from the theory without having been assumed at the outset.

If the theory is correct, space is not quite what it appears to be. In particular, the theory predicts that space has precisely nine dimensions (so spacetime has 10 dimensions once time is included), which represent six more than the usual three of length, breadth and height. Those extra dimensions are invisible to us. For instance, they may be very small and we may be oblivious to them simply because we cannot fit into them. A parking lot may have a hairline fracture, adding a third dimension (depth) to the pavement surface, but if the fracture is small, you will never notice it. Even string theorists have difficulty visualizing nine dimensions, but if the history of physics has taught us anything, it is that the true nature of the world may lie beyond our ability to visualize directly.

Despite its name, the theory is not just about strings. It also contains another kind of object called a Dirichlet brane—D-brane, for short. D-branes are large, massive surfaces that float within space. They act like slippery sheets of flypaper: the ends of open strings move on them but cannot be pulled off. Subatomic particles such as electrons and protons may be nothing more than open strings and, if so, are stuck to a brane. Only a few hypothetical particles, such as the graviton (which transmits the force of gravity), must be closed strings and are thus able to move completely freely through the extra dimensions. This distinction offers a second reason not to see the extra dimensions: our instruments may be built of particles that are trapped on a brane. If so, future instruments might be able to use gravitons to reach out into the extra dimensions.

D-branes can have any number of dimensions up to nine. A zero-dimensional D-brane (D0-brane) is a special type of particle, a D1-brane is a special type of string (not the same as a fundamental string), a D2-brane is a membrane or a wall, a D3-brane is a volume with height, depth and width, and so on. Our entire observed universe could be trapped on such a brane—a so-called brane world. Other brane worlds may float around out there, each being a universe to those trapped onboard. Because branes can move in the extra dimensions, they can behave like particles. They can move, collide, annihilate, and even form systems of branes orbiting around one another like planets.

Although these concepts are provocative, the acid test of a theory comes when it is confronted with experiments. Here string theory has disappointed because it has not yet been possible to test it experimentally, despite more than 20 years of continued investigation. It has proved hard to find a smoking gun—a prediction that, when tested, would decisively tell us whether or not the world is made of strings. Even the Large Hadron Collider (LHC)—which is now nearing completion at CERN, the European laboratory for particle physics near Geneva—may not be powerful enough.

Seeing the Unseen Dimensions

Which brings us back to inflation. If inflation occurs at the high energies where the stringy nature of particles becomes conspicuous, it may provide the very experimental tests that string theorists

Section 1: Multiple Dimensions

have been looking for. In the past few years, physicists have begun to investigate whether string theory could explain inflation. Unfortunately, this goal is easier to state than achieve.

To be more specific, physicists are checking whether string theory predicts a scalar field with two properties. First, its potential energy must be large, positive and roughly constant, so as to drive inflation with vigor. Second, this potential energy must be able to convert abruptly into kinetic energy—the exhilarating roller-coaster plunge at the end of inflation.

The good news is that string theory predicts no shortage of scalar fields. Such fields are a kind of consolation prize for creatures such as ourselves who are stuck in three dimensions: although we cannot peer into the extra dimensions, we perceive them indirectly as scalar fields. The situation is analogous to taking an airplane ride with all the window shades lowered. You cannot see the third dimension (altitude), but you can feel its effects when your ears pop. The change in pressure (a scalar field) is an indirect way of perceiving the dimension.

Air pressure represents the weight of the column of atmosphere above your head. What do the scalar fields of string theory represent? Some correspond to the size or shape of space in the unseen directions and are known by the mathematical term of geometric "moduli" fields. Others represent the distance between brane worlds. For instance, if our D3-brane approached another D3-brane, the distance between the two could vary slightly with location because of ripples in each brane. Physicists in Toronto might measure a scalar field value of 1 and physicists in Cambridge a value of 2, in which case they could conclude that the neighboring brane is twice as far from Cambridge as from Toronto.

To push two branes together or contort extra-dimensional space takes energy, which can be described by a scalar field. Such energy might cause branes to inflate, as first proposed by Georgi Dvali of New York University and Henry S.-H. Tye of Cornell University in 1998. The bad news is that the first calculations for the various scalar fields were not encouraging. Their energy density proved to be very low—too low to drive inflation. The energy profile more resembled a train sitting on level ground than a slowly climbing roller coaster.

Introducing Antibranes

That is where the problem stood when the two of us—together with Mahbub Majumdar, then at the University of Cambridge, and Govindan Rajesh, Ren-Jie Zhang and the late Detlef Nolte, all then at the Institute for Advanced Study in Princeton, N.J.—began thinking about it in 2001. Dvali, Sviatoslav Solganik of N.Y.U. and Qaisar Shafi of the University of Delaware developed a related approach at the same time.

Our innovation was to consider both branes and antibranes. Antibranes are to branes what antimatter is to matter. They attract each other, much as electrons attract their antiparticles (positrons). If a brane came near an antibrane, the two would pull each other together. The energy inside the branes could provide the positive energy needed to start inflation, and their mutual attraction could provide the reason for it to end, with the brane and antibrane colliding to annihilate each other in a grand explosion. Fortunately, our universe does not have to be annihilated to benefit from this inflationary process. When branes attract and annihilate, the effects spill over into nearby branes.

When we calculated the attractive force in this model, it was too strong to explain inflation, but the model was a proof of principle, showing how a steady process could have an abrupt ending that might fill our universe with particles. Our hypothesis of antibranes also inspired new thinking on the long-standing question of why our universe is three-dimensional.

The next level of refinement was to ask what happens when space itself, not just the branes within it, becomes dynamic. In our initial efforts, we had assumed the size and shape of extra-dimensional space to be fixed as the branes moved around. That was a serious omission, because space bends in response to matter, but an understandable one, because in 2001 nobody knew how to compute this extra-dimensional bending explicitly within string theory.

Space Warps

Within two years the situation changed dramatically. In 2003 a new theoretical framework known as KKLT, for its creators' initials, was developed by Shamit Kachru, Renata Kallosh and Andrei Linde of Stanford University, together with Sandip Trivedi of the Tata Institute of Fundamental Research in Mumbai. Their framework describes the circumstances in which the geometry of the extra dimensions is very stiff and so does not flex too much as things move around within it. It predicts a huge number of possible configurations for the extra dimensions, each corresponding to a different possible universe. The set of possibilities is known as the string theory landscape. Each possibility might be realized in its own region of the multiverse.

Within the KKLT framework, inflation can happen in at least two ways. First, it could result from the gravitational response of extra dimensions to brane-antibrane motion. The extra-dimensional geometry can be very peculiar, resembling an octopus with several elongations, or "throats." If a brane moves along one of these throats, its motion through the warped dimensions weakens the brane-antibrane attraction. This weakening enables the slow-roll process that gives rise to inflation, perhaps solving the main problem with our original proposal.

Second, inflation could be driven purely by changes in the geometry of the extra dimensions, without the need for mobile branes at all. Two years ago we and our colleagues presented the first stringy inflationary scenario along the second of these lines. This general process is called moduli inflation because moduli fields, which describe the geometry, act as the inflatons. As the extra dimensions settle into their current configuration, the three normal dimensions expand at an accelerated pace. In essence, the universe sculpts itself. Moduli inflation thus relates the size of the dimensions we see to the size and shape of those we cannot.

Strings in the Sky

The stringy inflation models, unlike many other aspects of string theory, might be tested observationally in the near future. Cosmologists have long thought that inflation would produce gravitational waves, ripples in the fabric of space and time. String theory may alter this prediction, because the existing stringy inflation models predict unobservably weak gravitational waves. The Planck satellite will be more sensitive to primordial gravitational waves than current instruments are. If it were to detect such waves, it would rule out all the models of string inflation proposed so far.

Also, some brane inflation models predict large linear structures known as cosmic strings, which naturally arise in the aftermath of brane-antibrane annihilation. These strings could come in several types: D1-branes or fundamental strings blown up to enormous size, or a combination of the two. If they exist, astronomers should be able to detect them by the way they distort the light coming from galaxies.

Despite the theoretical progress, many open questions remain. Whether inflation indeed occurred is not entirely settled. If improved observations cast doubt on it, cosmologists will have to turn to alternative pictures of the very early universe. String theory has inspired several such alternatives, in which our universe existed before the big bang, perhaps as part of a perpetual cycle of creation and destruction. The difficulty in these cases is to describe properly the transition that marks the moment of the big bang.

In summary, string theory provides two general mechanisms for obtaining cosmic inflation: the collision of branes and the reshaping of extra-dimensional spacetime. For the first time, physicists have been able to derive concrete models of cosmic inflation rather than being forced to make uncontrolled, ad hoc assumptions. The progress is very encouraging. String theory, born of efforts to explain phenomena at minuscule scales, may be writ large across the sky.

About the Authors

Cliff Burgess and Fernando Quevedo met in the early 1980s as graduate students of the famous physicist Steven Weinberg. They have worked together ever since, mostly on the question of how to relate string theory to real live observable physics. Burgess is a researcher at the Perimeter Institute in Waterloo, Ontario, and a professor at McMaster University in nearby Hamilton. He received a Killam Fellowship in 2005. Quevedo is a professor at the University of Cambridge and a recipient of a Guggenheim Fellowship, among other awards. He has been active in developing science in his home country of Guatemala.

Long Live the Multiverse!

By Tom Siegfried

Ernst Mach, the Austrian physicist-philosopher of the late 19th century, famously denied the reality of atoms. "Have you ever seen one?" he mockingly asked of atom advocates. Today many scientists speak with similar derision about the idea that the visible universe is not alone, but rather is only one of many universes—a single bubble in a froth of cosmic carbonation known as the multiverse.

You can't see these other universes, so the idea is not testable, multiverse opponents allege. Besides, invoking a multiplicity of universes to explain reality is a violent violation of Occam's razor, the philosophical principle favoring simple explanations over complicated ones.

But Mach, of course, was wrong about atoms. And throughout history, those arguing against multiple universes have invariably turned out to be wrong as well. In fact, the first proponents of the multiverse were the same ancient Greeks who proposed the existence of atoms. Leucippus and Democritus believed that their atomic theory required an infinity of worlds ("world" being synonymous with "universe"). Their later follower, Epicurus of Samos, also professed the reality of multiple worlds. "There are infinite worlds both like and unlike this world of ours," he averred.

Aristotle, however, argued strongly that logic required one universe only. His view prevailed until 1277, when the bishop of Paris declared that medieval scholars teaching Aristotle's view would be excommunicated—for denying God's power to create as many universes as he wanted to. Centuries of debate followed. Some argued that God could create more universes but probably didn't; others maintained that reality comprised a "plurality of worlds."

In the 16th century, Copernicus turned the issue on its head. Instead of Aristotle's universe (Earth in the middle, surrounded by planets affixed to rotating spheres), Copernicus placed the sun

Section 1: Multiple Dimensions

in the middle, with the planets (including Earth) in orbit. The universe became a solar system, bounded by a sphere of stars. Shortly thereafter Thomas Digges in England redrew the Copernican picture, with stars littered throughout distant space rather than fixed to a single sphere. That raised the possibility of multiple solar system universes scattered throughout the heavens. Giordano Bruno, perhaps influenced by Digges, proclaimed that God is glorified "not in one, but in countless suns; not in a single earth, a single world, but in a thousand thousand, I say in an infinity of worlds."

Bruno's contemporary, the famed astronomer Johannes Kepler, didn't like that idea. He conceived the universe as the solar system. Similar worlds beyond our sight are not scientific. "If they are not seen," Kepler declared, "they for this reason are not pertinent to astronomy." Anything beyond what's visible, he insisted, "is superfluous metaphysics"—a view strikingly similar to the attitude of many toward the multiverse today.

Kepler was wrong, of course. Later telescopes revealed a multitude of stars at great distances, congregating in a lens-like disk, the Milky Way galaxy (of which the sun was one member). Just as Copernicus showed that the Earth is part of a solar system universe, the solar system became just one of many such "universes" in the Milky Way. Once again, the universe was redefined—no longer a set of spheres surrounding the Earth, or a set of planets orbiting the sun, but now a vast disk of stars surrounded by emptiness.

Except in that emptiness appeared fuzzy blobs, called nebulae. Immanuel Kant and others speculated that those blobs were actually galaxies themselves, just very far away—island universes, to use the term coined in the 1840s by the American astronomer Ormsby MacKnight Mitchel. This new vision of a multiverse also met with ridicule. "No competent thinker" believed in island universes, the astronomy writer Agnes Clerke declared at the end of the 19th century. It was an idea that had withdrawn "into the region of discarded and half-forgotten speculations."

But once again, the multiverse prevailed. In 1924 Edwin Hubble reported proof that some of those fuzzy nebulae, such as

Is There More Than One Universe?

Andromeda, were indeed island universes as grand as the Milky Way. Hubble pioneered today's current definition of the universe as a vast expanding bubble of spacetime populated by billions and billions of such galaxies.

In the 1980s, a new explanation for how that universe came to be, called inflationary cosmology, revived the multiverse question in a novel way. If the initial big bang launching our universe into existence was followed by a burst of extremely rapid expansion (inflation), that same inflationary event could have recurred in other parts of space. If inflation theory turns out to be correct, our bubble would then be only one of many.

Of course, just because multiverse advocates have been right historically doesn't mean that they will certainly be right again this time. But multiverse opponents are certainly wrong to say that the multiverse idea is not science because it is not testable. The multiverse is not a theory to be tested, but rather a prediction of other theories that can be tested. Inflationary cosmology has, in fact, already passed many tests, although not yet enough to be definitively established.

For that matter, it's not necessarily true that other universes are in principle not observable. If another bubble collided with ours, telltale marks might appear in the cosmic background radiation left over from the big bang. Even without such direct evidence, their presence might be inferred by indirect means, just as Einstein demonstrated the existence of atoms in 1905 by analyzing the random motion of particles suspended in liquid.

Today, atoms actually can be "seen," in images produced by scanning tunneling microscopes. Atoms did not suddenly become real when first imaged, though; they had been legitimate scientific entities for two and a half millennia. Multiple universes have been a topic of philosophical-scientific discussion for just as long.

As for Occam's razor, you could check with William of Occam himself, the 14th-century philosopher who articulated that principle. In his day, he was the most enthusiastic of the advocates for a multiplicity of worlds.

Section 1: Multiple Dimensions

The views expressed are those of the author(s) and are not necessarily those of Scientific American.

About the Author

Tom Siegfried is a science writer and editor in the Washington, DC, area. His latest book, The Number of the Heavens, *about the history of the multiverse, was published in September by Harvard University Press.*

Our Improbable Existence Is No Evidence for a Multiverse

By Philip Goff

We exist, and we are living creatures. It follows that the universe we live in must be compatible with the existence of life. However, as scientists have studied the fundamental principles that govern our universe, they have discovered that the odds of a universe like ours being compatible with life are astronomically low. We can model what the universe would have looked like if its constants—the strength of gravity, the mass of an electron, the cosmological constant—had been slightly different. What has become clear is that, across a huge range of these constants, they had to have pretty much exactly the values they had in order for life to be possible. The physicist Lee Smolin has calculated that the odds of life-compatible numbers coming up by chance is 1 in 10^{229}.

Physicists refer to this discovery as the "fine-tuning" of physics for life. What should we make of it? Some take this to be evidence of nothing other than our good fortune. But many prominent scientists—Martin Rees, Alan Guth, Max Tegmark—have taken it to be evidence that we live in a multiverse: that our universe is just one of a huge, perhaps infinite, ensemble of worlds. The hope is that this allows us to give a "monkeys on typewriters" explanation of the fine-tuning. If you have enough monkeys randomly jabbing away on typewriters, it becomes not so improbable that one will happen to write a bit of English. By analogy, if there are enough universes, with enough variation in the numbers in their physics, then it becomes statistically likely that one will happen to have the right numbers for life.

This explanation makes intuitive sense. However, experts in the mathematics of probability have identified the inference from the fine-tuning to the multiverse as an instance of fallacious reasoning. Specifically, multiverse theorists commit the inverse gambler's fallacy, which is a slight twist on the regular gambler's fallacy. In

Section 1: Multiple Dimensions

the regular gambler's fallacy, the gambler has been at the casino all night and has had a terrible run of bad luck. She thinks to herself, "My next roll of the dice is bound to be a good one, as it's unlikely I'd roll badly all night!" This is a fallacy, because for any particular roll, the odds of, say, getting a double six are the same: 1/36. How many times the gambler has rolled that night has no bearing on whether the *next roll* will be a double six.

In the inverse gambler's fallacy, a visitor walks into a casino and the first thing she sees is someone rolling a double six. She thinks "Wow, that person must've been playing for a long time, as it's unlikely they'd have such good luck just from one roll." This is fallacious for the same reason. The casino-visitor has only observed one roll of the dice, and the odds of that one roll coming good is the same as any other roll: 1/36. How long the player has been rolling prior to this moment has no bearing on the odds of the one roll the visitor observed being a double six.

Philosopher Ian Hacking was the first to connect the inverse gambler's fallacy to arguments for the multiverse, focusing on physicist John Wheeler's oscillating universe theory, which held that our universe is the latest of a long temporal sequence of universes. Just as the casino-visitor says "Wow, that person must've been playing for a long time, as it's unlikely they'd have such good luck just from one roll," so the multiverse theorist says "Wow, there must be many other universes before this one, as it's unlikely the right numbers would have come up if there'd only been one."

Other theorists later realized that the charge applies quite generally to every attempt to derive a multiverse from fine-tuning. Consider the following analogy. You wake up with amnesia, with no clue as to how you got where you are. In front of you is a monkey bashing away on a typewriter, writing perfect English. This clearly requires explanation. You might think: "Maybe I'm dreaming ... maybe this is a trained monkey ... maybe it's a robot." What you would not think is "There must be lots of other monkeys around here, mostly writing nonsense." You wouldn't think this because what needs explaining is why *this* monkey—the only one you've

Is There More Than One Universe?

actually observed—is writing English, and postulating other monkeys doesn't explain what *this* monkey is doing.

Some have objected that this argument against the inference from fine-tuning to a multiverse ignores the *selection effect* that exist in cases of fine-tuning, namely that fact that we could not possibly have observed a universe that wasn't fine-tuned. If the universe wasn't fine-tuned, then life would be impossible, and so nobody would be around to observe anything. It is of course true that this selection effect exists, but it makes no difference to whether or not the fallacy is committed. We can see this by just adding an artificial selection effect to the monkey and typewriter analogy of the last paragraph. Consider the following story:

You wake up to find yourself in a room sat opposite the Joker (from *Batman*) and a monkey called Joey on a typewriter. The Joker tells you that while you were unconscious, he decided to play a little game. He gave Joey one hour to bash on the typewriter, committing to release you if Joey wrote some English or to kill you before you regained consciousness if he didn't. Fortunately, Joey has typed "I love how yellow bananas are," and hence you are to be released.

In the above story, you could not possibly have observed Joey typing anything other than English—the Joker would have killed you before you had a chance—just as we could never have observed a non-fine-tuned universe. And yet the inference to many monkeys is still unwarranted. Given how unlikely it is that an ordinary monkey would come up with "I love how yellow bananas are" just by randomly bashing away, you might suspect some kind of trick. What you would not conclude, however, is that there must be many other monkeys typing rubbish. Again, what you need explaining is why *Joey* is typing English, and the postulation of other monkeys doesn't explain this. By analogy, what we need explaining is why the only universe we've ever observed is fine-tuned, and the postulation of other universes doesn't account for this.

But isn't there scientific evidence for a multiverse? Some physicists do indeed think there is a tentative empirical evidence for a kind of multiverse, that described by the hypothesis of eternal inflation. According to eternal inflation, there is a vast, exponentially expanding mega space in which certain regions slow down to form "bubble universes," our universe being one such bubble universe. However, there is no empirical ground for thinking that the constants of physics—the strength of gravity, the mass of electrons, etc.—are different in these different bubble universes. And without such variation, the fine-tuning problem is even worse: we now have a huge number of monkeys *all* of whom are typing English.

At this point, many bring in string theory. String theory offers a way to make sense of the *possibility* that the different bubbles might have different constants. On string theory, the supposedly "fixed" numbers of physics are determined by the phase of space, and there are 10^{500} different possible phases of space in the so-called "string landscape." It could be that random processes ensure that a wide variety of possibilities from the string landscape are realized in the different bubble universes. Again, however, there is no empirical reason for thinking that this possibility is actual.

The reason some scientists take seriously the possibility of a multiverse in which the constants vary in different universes is that it seems to explain the fine-tuning. But on closer examination, the inference from fine-tuning to the multiverse proves to be instance of flawed reasoning. So, what should we make of the fine-tuning? Perhaps there is some other way of explaining it. Or perhaps we just got lucky.

About the Author

Philip Goff is a philosopher and consciousness researcher at Durham University in the UK, and author of Galileo's Error: Foundations for a New Science of Consciousness. *His research focuses on how to integrate consciousness into our scientific worldview. His website is www.philipgoffphilosophy.com and he blogs at https://conscienceandconsciousness.com.*

Extra Dimensions

By George Musser

Wouldn't it be great to reach your arm into a fourth dimension of space? You could then liberate yourself from the shackles of ordinary geometry. Hopelessly tangled extension cords would slip apart with ease. A left-handed glove could be flipped over to replace the right-handed one your dog ate. Dentists could do root canals without drilling or even asking you to open your mouth.

As fantastic as extra dimensions of space sound, they might really exist. From the relative weakness of gravity to the deep affinity among seemingly distinct particles and forces, various mysteries of the world around us give the impression that the known universe is but the shadow of a higher-dimensional reality. If so, the Large Hadron Collider (LHC) near Geneva could smash particles together and release enough energy to break the shackles that keep particles in three dimensions and let us reach into that mind-blowing realm.

Proof of extra dimensions "would alter our whole notion of what reality is," says cosmologist Max Tegmark of the Massachusetts Institute of Technology, who in 1990 wrote a four-dimensional version of the video game Tetris to get a taste of what extra dimensions might be like. (You keep track of the falling blocks using multiple 3-D slices of the full 4-D space.)

In modern physics theories, the main rationale for extra dimensions is the concept of supersymmetry, which aims to unite all the different types of particles into one big happy family. Supersymmetry can fulfill that promise only if space has a total of 10 dimensions. The dimensions could have gone unnoticed either because they are too small to enter or because we are, by our very nature, stuck to a 3-D membrane like a caterpillar clutching onto a leaf.

To be sure, not every proposed unified theory involves extra dimensions. So their discovery or nondiscovery would be a helpful data point. "It would focus what we do," says physicist Lisa Randall

Section 1: Multiple Dimensions

of Harvard University, who made her name studying the caterpillar-and-leaf option.

One way to get at those dimensions is to crank up the energy of a particle accelerator. By the laws of quantum mechanics, the more energy a particle has, the more tightly confined it is; an energy of one tera-electron-volt (TeV) corresponds to a size of 1019 meter. If an extra dimension is that big, the particle would literally fall into it and begin to vibrate.

In 1998 physicist Gordon Kane of the University of Michigan at Ann Arbor imagined that the LHC smashed together two protons and created electrons and other particles that not only had the energy of 1 TeV but also integer multiples thereof, such as 2 or 3 TeV. Such multiples would represent the harmonics of the vibrations in extra dimensions set off by the collision. Neither standard particle processes nor exotica such as dark matter particles could account for these events.

Extra dimensions might betray themselves in other ways. If the LHC produced subatomic black holes, they would be immediate proof of extra dimensions, because gravity in ordinary 3-D space is simply too weak to create holes of this size. For geometric reasons, higher dimensions would strengthen gravity on small scales. They would likewise change the small-scale behavior of other forces, such as electromagnetism. And by dictating how supersymmetry operates, they might lead to distinctive patterns among the masses and other properties of particles. Besides the LHC, scientists might find hints of extra dimensions in measurements of the strength of gravity and in observations of the orbits of black holes or of exploding stars.

The discovery would transform not only physics but also its allied disciplines. Extra dimensions might explain mysteries such as cosmic acceleration and might even be a prelude to reworking the entire notion of dimensionality—adding to a growing sense that space and time emerge from physical principles that play out in a spaceless, timeless realm.

"So while extra dimensions would be a terrific discovery," says physicist Nima Arkani-Hamed of the Institute for Advanced Study in Princeton, N.J., "at a deeper level, conceptually they aren't particularly fundamental."

Whatever the charms of extra dimensions for physicists, we will never be able to visit them for ourselves. If they were open to the particles that make up our bodies, the added liberty of motion would destabilize complex structures, including life. Alas, the frustration of tangled cords and the pain of dental work are necessary trade-offs to allow us to exist at all.

About the Author

George Musser is a contributing editor at Scientific American *and author of* Spooky Action at a Distance *(Farrar, Straus and Giroux, 2015) and* The Complete Idiot's Guide to String Theory *(Alpha, 2008).*

New Phase of Matter Opens Portal to Extra Time Dimension

By Zeeya Merali

When the ancient Incas wanted to archive tax and census records, they used a device made up of a number of strings called a *quipu*, which encoded the data in knots. Fast-forward several hundred years, and physicists are on their way to developing a far more sophisticated modern equivalent. Their *"quipu"* is a new phase of matter created within a quantum computer, their strings are atoms, and the knots are generated by patterns of laser pulses that effectively open up a second dimension of time.

This isn't quite as incomprehensible as it first appears. The new phase is one of many within a family of so-called topological phases, which were first identified in the 1980s. These materials display order not on the basis of how their constituents are arranged—like the regular spacing of atoms in a crystal—but on their dynamic motions and interactions. Creating a new topological phase—that is, a new "phase of matter"—is as simple as applying novel combinations of electromagnetic fields and laser pulses to bring order or "symmetry" to the motions and states of a substance's atoms. Such symmetries can exist in time rather than space, for example in induced repetitive motions. Time symmetries can be difficult to see directly but can be revealed mathematically by imagining the real-world material as a lower-dimensional projection from a hypothetical higher-dimensional space, similar to how a two-dimensional hologram is a lower-dimensional projection of a three-dimensional object. In the case of this newly created phase, which manifests in a strand of ions (electrically charged atoms), its symmetries can be discerned by considering it as a material that exists in higher-dimensional reality with two time dimensions.

"It is very exciting to see this unusual phase of matter realized in an actual experiment, especially because the mathematical description is based on a theoretical 'extra' time dimension," says team member Philipp Dumitrescu, who was at the Flatiron Institute in New York City when the experiments were carried out. A paper describing the work was published in *Nature* on July 20.

Opening a portal to an extra time dimension—even just a theoretical one—sounds thrilling, but it was not the physicists' original plan. "We were very much motivated to see what new types of phases could be created," says study co-author Andrew Potter, a quantum physicist at the University of British Columbia. Only after envisioning their proposed new phase did the team members realize it could help protect data being processed in quantum computers from errors.

Standard classical computers encode information as strings of bits—0's or 1's—while the predicted power of quantum computers derives from the ability of quantum bits, or qubits, to store values of either 0 or 1, or both simultaneously (think Schrödinger's cat, which can be both dead and alive). Most quantum computers encode information in the state of each qubit, for instance in an internal quantum property of a particle called spin, which can point up or down, corresponding to a 0 or 1, or both at the same time. But any noise—a stray magnetic field, say—could wreak havoc on a carefully prepared system by flipping spins willy-nilly and even destroying quantum effects entirely, thereby halting calculations.

Potter likens this vulnerability to conveying a message using pieces of string, with each string arranged in the shape of an individual letter and laid out on the floor. "You could read it fine until a small breeze comes along and blows a letter away," he says. To create the more error-proof quantum material, Potter's team looked to topological phases. In a quantum computer that exploits topology, information is not encoded locally in the state of each qubit but is woven across the material globally. "It's like a knot that's hard to undo—like quipu," the Incas' mechanism for storing census and other data, Potter says.

"Topological phases are intriguing because they offer a way to protect against errors that's built into the material," adds study

Section 1: Multiple Dimensions

co-author Justin Bohnet, a quantum physicist at the company Quantinuum in Broomfield, Colo., where the experiments were carried out. "This is different to traditional error-correcting protocols, where you are constantly doing measurements on a small piece of the system to check if errors are there and then going in and correcting them."

Quantinuum's H1 quantum processor is made up of a strand of 10 qubits—10 ytterbium ions—in a vacuum chamber, with lasers tightly controlling their positions and states. Such an "ion trap" is a standard technique used by physicists to manipulate ions. In their first attempt to create a topological phase that would be stable against errors, Potter, Dumitrescu and their colleagues sought to imbue the processor with a simple time symmetry by imparting periodic kicks to the ions—all lined up in one dimension—with regularly repeating laser pulses. "Our back-of-the-envelope calculations suggested this would protect [the quantum processor] from errors," Potter says. This is similar to how a steady drumbeat can keep multiple dancers in rhythm.

To see if they were right, the researchers ran the program multiple times on Quantinuum's processor and checked each time to see if the resulting quantum state of all the qubits matched their theoretical predictions. "It didn't work at all," Potter says with a laugh. "Totally incomprehensible stuff was coming out." Each time, accumulating errors in the system degraded its performance within 1.5 seconds. The team soon realized that it was not enough to just add one time symmetry. In fact, rather than preventing the qubits from being affected by outside knocks and noise, the periodic laser pulses were amplifying tiny hiccups in the system, making small disruptions even worse, Potter explains.

So he and his colleagues went back to the drawing board until, at last, they struck upon an insight: if they could concoct a pattern of pulses that was somehow itself ordered (rather than random) yet did not repeat in a regular manner, they might create a more resilient topological phase. They calculated that such a "quasi-periodic" pattern could potentially induce multiple symmetries

Is There More Than One Universe?

in the processor's ytterbium qubits while avoiding the unwanted amplifications. The pattern they chose was the mathematically well-studied Fibonacci sequence, in which the next number in the sequence is the sum of the previous two. (So where a regular periodic laser pulse sequence might alternate between two frequencies from two lasers as A, B, A, B..., a pulsing Fibonacci sequence would run as A, AB, ABA, ABAAB, ABAABABA....)

Although these patterns actually emerged from a rather complex arrangement of two collections of varying laser pulses, the system, according to Potter, can be simply considered as "two lasers pulsing with two different frequencies" that ensure the pulses never temporally overlap. For the purpose of its calculations, the theoretical side of the team imagined these two independent collections of beats along two separate time lines; each collection is effectively pulsing in its own time dimension. These two time dimensions can be traced on to the surface of a torus. The quasi-periodic nature of the dual time lines becomes clear by the way they each wrap around the torus again and again "at a weird angle that never repeats on itself," Potter says.

When the team implemented the new program with the quasi-periodic sequence, Quantinuum's processor was indeed protected for the full length of the test: 5.5 seconds. "It doesn't sound like a lot in seconds, but it's a really stark difference," Bohnet says. "It's a clear sign the demonstration is working."

"It's pretty cool," agrees Chetan Nayak, an expert on quantum computing at Microsoft Station Q at the University of California, Santa Barbara, who was not involved in the study. He notes that, in general, two-dimensional spatial systems offer better protection against errors than one-dimensional systems do, but they are harder and more expensive to build. The effective second time dimension created by the team sneaks round this limitation. "Their one-dimensional system acts like a higher-dimensional system in some ways but without the overhead of making a two-dimensional system," he says. "It's the best of both worlds, so you have your cake and you eat it, too."

Samuli Autti, a quantum physicist at Lancaster University in England, who was also not involved with the team, describes the tests as "elegant" and "fascinating" and is particularly impressed that they involve "dynamics"—that is, the laser pulses and manipulations that stabilize the system and move its constituent qubits. Most previous efforts to topologically boost quantum computers have relied on less active control methods, making them more static and less flexible. Thus, Autti says, "Dynamics with topological protection is a major technological goal."

The name the researchers assigned to their new topological phase of matter recognizes its potentially transformative capabilities, although it is a bit of a mouthful: emergent dynamical symmetry-protected topological phase, or EDSPT. "It'd be nice to think of a catchier name," Potter admits.

There was another unexpected bonus of the project: the original failed test with the periodic pulse sequence revealed that the quantum computer was more error-prone than assumed. "This was a good way of stretching and testing how good Quantinuum's processor is," Nayak says.

About the Author

Zeeya Merali is a freelance writer based in London and author of A Big Bang in a Little Room.

Section 2: Time and Space

2.1 How Scientists Solved One of the Greatest Open Questions in Quantum Physics
By Spyridon Michalakis

2.2 Quantum Gravity in Flatland
By Steven Carlip

2.3 The Beauty at the Heart of a "Spooky" Mystery
By John Horgan

2.4 This Twist on Schrödinger's Cat Paradox Has Major Implications for Quantum Theory
By Zeeya Merali

2.5 Can Quantum Mechanics Save the Cosmic Multiverse?
By Yasunori Nomura

2.6 The Difficult Birth of the "Many Worlds" Interpretation of Quantum Mechanics
By Adam Becker

2.7 Black Hole Discovery Helps to Explain Quantum Nature of the Cosmos
By Edgar Shaghoulian

2.8 How the Inside of a Black Hole Is Secretly on the Outside
By Ahmed Almheiri

How Scientists Solved One of the Greatest Open Questions in Quantum Physics

By Spyridon Michalakis

I am sitting alone at the head of a large conference table when an oddly familiar voice greets me: "Hey, you must be Spiros!" I turn around to find Paul Rudd, the Hollywood actor, wearing his famed disarming smile. He is in sweats, on his way back from some type of superhero training.

A few minutes later he and a bunch of other film people are sitting around me. Rudd cuts straight to the chase: "So what kinds of cool things happen when you shrink?" I have been flown in to consult on the physics of Marvel Studios' superhero flick *Ant-Man*, and now I must deliver. Yet all I really know about shrinking to ant size comes from watching *Honey, I Shrunk the Kids!* as a nine-year-old. For a moment, I consider telling him that he's got the wrong guy, but there is no way I am going to let this opportunity slip between my fingers. I may not know much about ants, but I know a thing or two about quantum physics. "The concepts of time and space lose their usual meaning when you shrink to the quantum scale," I reply with confidence. Reading the room, I can tell that this is the last thing they expected to hear. But they are hooked. The floor is mine for the next two hours, as I delve deeper and deeper into the rules and weirdness of quantum mechanics.

A day later one of the producers e-mails me: "Hey, what should we call the place you enter when you shrink to microscopic size?" I type back: "How about the Quantum Realm?" Five years later, in 2019, Marvel's Avengers enter the Quantum Realm and travel back in time to save the universe. All of a sudden, being an expert in quantum physics seems pretty cool.

I was not always into physics or comic-book heroes. In college, I majored in mathematics and computer science, spending my summers trying to predict how one-dimensional DNA sequences folded into three-dimensional proteins. It was not until graduate school that I took my first physics class beyond the basic college requirements. My Ph.D. adviser at the University of California, Davis, had decided to enroll me in graduate-level quantum mechanics, and I had no choice but to go along with it. When on the first day of class we were handed a one-page undergraduate-level assessment test, I returned mine with my name and a smiley face next to it. Still, I persisted, graduating in June 2008 with a doctorate in applied mathematics and an emphasis on mathematical physics and quantum information theory. Three months later I would pack my things and move to Los Alamos, N.M., the birthplace of the atomic bomb, to take a postdoctoral position at Los Alamos National Laboratory. I did not know it at the time, but during the next year I would delve deep within the quantum realm. This is the story of what I discovered there and how I made it back to tell Marvel the story.

Something Interesting

It all began with a simple question.

My adviser at Los Alamos, Matthew Hastings, a rising star and one of the sharpest minds in physics, was sitting across from me at a sushi restaurant when he popped the fateful question: "For your postdoc here at the lab, do you want to start with a warm-up, or do you want to work on something interesting?" Without asking for further clarification, I answered, "I want to work on something interesting." He seemed pleased with my answer. Later that day he sent me a link to a list of 13 unsolved problems in physics maintained by Michael Aizenman, a professor at Princeton University and a towering figure in mathematical physics. I was to work on the second problem on that list, a question posed by mathematical physicists Joseph Avron and Ruedi Seiler: "Why is the Hall conductance quantized?"

Section 2: Time and Space

You may wonder what the Hall conductance is or what it means for it to be quantized. I had the same questions back then. No problem on the list besides the third—cryptically titled "Exponents and Dimensions"—had "SOLVED!" next to it. Clicking through, I saw that it was actually only partially solved. Yet one of those partial breakthroughs had led to a Fields Medal, one of the highest honors in mathematics, in 2006, and the other would earn one four years later. In this company, it was clear the problem I was tasked with solving was no ordinary quandary. I considered carefully if I could solve such a question within a year. The reason for the time limit is that a postdoc in math or physics usually lasts two years. At the end of your first year, if you have done great research, you may apply to top universities for a tenure-track professorship. If your research is good but not great, you may apply for a second postdoc or look for a less competitive tenure-track position. If you have nothing to show after your first year, there is always Wall Street.

Still, the idea of backing out now, without even trying to attack the problem, was difficult to swallow. For a person growing up in Spata, a small town outside of Athens, Greece, big dreams were unusual. My dad grew up in the same house I did. He played soccer and got into fights. When he eventually dropped out of high school, his dad offered him a position at the local grocery store. My father refused. Despite being a dropout, he had ambition. He interned at the local real-estate agency and learned the ropes of buying and selling land. Later, he went back to school to get his GED at my mother's insistence. Down the line, when my older brother, Nikos, brought home his first-grade report card, my father cried with happiness when he realized that his son was a good student. Nikos and I would go on to compete at the International Mathematical Olympiad, an honor afforded to six high school students from each country every year. Then, one after the other, Nikos, I and my younger brother, Marios, traded high school in Athens for college at the Massachusetts Institute of Technology in Cambridge—a rare accomplishment for any family, let alone one of modest means, and a testament to my parents. I thought that if they could perform miracles, maybe I could, too. So, in the fall of 2008, I began working on problem number two, aiming, as the list put it, to "formulate the theory of the integer quantum Hall effect, which explains the quantization of the Hall conductance, so that it applies also for interacting electrons in the thermodynamic limit."

The integer quantum Hall effect has a long history. The original Hall effect was discovered in 1879 by Edwin H. Hall, a student at Johns Hopkins University. Young Hall had decided to challenge a claim made by the father of electromagnetism, James Clerk Maxwell. In his 1873 *Treatise on Electricity and Magnetism*, Maxwell confidently declared that, in the presence of a magnetic field, a conducting material with current flowing through it will bend because of the magnetic force on the material, not on the current. Maxwell concluded that "when a constant magnetic force is made to act on the system ... the distribution of the current will be found to be the same as if no magnetic force were in action." To test the idea, Hall ran current across a thin leaf of gold placed in a magnetic field perpendicular to its surface and noticed that his galvanometer (an instrument used to detect small currents) registered a current, which implied a voltage (electric potential) in a direction *perpendicular* to that of the current's original path. He concluded that the magnetic field was pushing the electrons in the current toward one edge of the conductor, permanently changing their distribution on the surface of the material. Maxwell was wrong. This unexpected charge buildup along the conductor's edges became known as the Hall voltage.

The *quantum* Hall effect was first observed nearly a century later, on February 5, 1980, in Grenoble, France, by German experimental physicist Klaus von Klitzing. His aim was to study the Hall effect more carefully under ultralow temperatures and high magnetic fields. He was looking for small deviations from the expected effect in certain two-dimensional semiconductors, the materials underlying all modern transistors. In particular, he was trying to measure the

Section 2: Time and Space

Hall resistance, a quantity proportional to the Hall voltage. What he observed was astonishing: the Hall resistance was quantized! Let me explain. As the strength of the magnetic field increased, the resistance between the edges of the material would stay exactly the same, until the field got high enough. Then, the resistance would jump to a new value instead of climbing up steadily the way Hall had originally observed—and all known physics at the time predicted. Even more surprisingly, the values of the Hall conductance, the inverse of the Hall resistance, were precise integer multiples of a quantity intimately related to the fine-structure constant, a fundamental constant of nature that describes the strength of the electromagnetic interaction between elementary charged particles. The integer quantum Hall effect was born.

Von Klitzing's discovery was remarkable, not least of all because the fine-structure constant was supposed to describe aspects of the quantum realm that were too fine-grained for any macroscopic phenomenon, such as the Hall conductance, to be able to probe, let alone define with incredible precision. Yet not only did the Hall conductance capture an essential aspect of the microscopic world of quantum physics, it did so with impossible ease. The integer plateaus of the Hall resistance appeared irrespective of variations in the size, the purity or even the particular type of semiconducting material used in the experiment. It was as if a symphony of a trillion trillion electrons maintained their collective quantum tune across vast atomic distances without the need for a master conductor and, even more astonishingly, were impervious to the principles of physics that, for billions of years, had guarded the quantum realm from macroscopic interlopers.

A door to the quantum realm was opened that day—a macroscopic door that many thought did not exist. In 1985, five years after the discovery, von Klitzing was awarded the Nobel Prize in Physics. His finding would lead to further breakthroughs, with three more Nobel Prizes awarded to two experimentalists (Horst Störmer and Daniel Tsui) and a theorist (Robert Laughlin) in 1998, for discovering that electrons acting together in strong magnetic fields can form new types of "particles," with charges that are mere fractions of electron charges, a phenomenon now known as the fractional quantum Hall effect.

Laughlin's Quantum Pump

Laughlin was one of the first physicists to attempt an explanation of the quantum Hall effect. In 1981 he came up with a brilliant thought experiment—an idealized simulation of the original experiment that provided a mathematical metaphor to understand it. Laughlin imagined electrons traveling along a conducting loop with a flat edge, like a wedding band. A magnetic field ran perpendicular to the surface of the band, but Laughlin added a fictitious magnetic field line—called a magnetic flux—threading through the middle of the loop like a finger through the ring. Increasing the fictional flux induced a current running around the loop, thus introducing the longitudinal current present in the classical Hall effect. The process, named Laughlin's quantum pump, would complete one cycle every time the fictional magnetic flux increased by one "flux quantum"—an amount defined as h/e, where h is Planck's constant and e is the electron's charge.

After each cycle, the quantum system would return to its original state as the result of a phenomenon known as gauge invariance. Laughlin argued that this reset implied that the Hall conductance was quantized in whole numbers equal to the number of electrons moved by the quantum pump. Great! Alas, there was an issue. The Hall conductance was experimentally measured (and averaged) over many cycles of the pump. Because Laughlin assumed (correctly) that the system was described by quantum mechanics, there was no guarantee that each cycle would transfer the same number of electrons. As Avron and Seiler would write later with their collaborator Daniel Osadchy: "Only in classical mechanics does an exact reproduction of a prior state guarantee reproduction of the prior measured result. In quantum mechanics, reproducing the state of the system does not necessarily reproduce the measurement outcome. So one cannot conclude from gauge invariance alone that the same number of electrons is transferred in every cycle of the pump." Physicists needed a new set of ideas to show that the average number of electrons transferred over several cycles was also an integer.

Inspired by Laughlin's argument, the next attempts at explaining the quantization of the Hall conductance relied heavily on the concept of adiabatic evolution. Adiabatic evolution is a physical process that aims to capture the evolution of a system that remains in its lowest-energy state at all times while some external parameter varies. When the system's spectral gap—the energy required for it to jump to an excited state—becomes small, adiabatic evolution slows down to prevent the system from crossing over to an excited state. Laughlin's original argument used this notion to mathematically model the quantum Hall effect as the adiabatic evolution of the electronic state of a quantum Hall system under the increase of a fictitious magnetic flux.

Unbreakable Play-Doh

To study the quantum Hall effect more deeply, physicists turned to a branch of mathematics called topology. Topology is a way of thinking about the fundamental essence of shapes—the properties that do not change even as they are continuously deformed. Think of a kind of Play-Doh that is unbreakable and impossible to glue onto itself. You can turn a cube of this substance into a ball by rounding out its sharp edges and corners, but you cannot turn it into a doughnut. The latter transformation would require either poking a hole through the cube or stretching and gluing it onto itself. In that sense, cubes and doughnuts are topologically distinct shapes, but cubes and balls are topologically the same (although they are all geometrically different). Topology was formalized in 1895 but had rarely interacted with physics until the 1950s and 1960s.

The initial efforts to understand the role of topology in the quantum Hall effect were considered so significant, in fact, that in 2016 theoretical physicists David Thouless and F. Duncan M. Haldane won a Nobel Prize for this work. Thouless and his collaborators, in particular, extended Laughlin's argument by showing that the Hall conductance was quantized on average. Because one fictitious flux was not enough to prove quantization, they proposed a second fictitious flux. In the new thought experiment, one flux induced the electric current across a semiconductor, and the other detected changes in the current between pump cycles. This scenario simulated cycles of Laughlin's pump under distinct initial conditions. The adiabatic evolution generated by the extra fictitious flux played the role of averaging over many cycles of Laughlin's pump and showed that the average Hall conductance was quantized.

At around the same time, Barry Simon, a mathematical physicist at the California Institute of Technology, noticed that adiabatic evolution formed a mathematical bridge between the Hall conductance and the local curvature of the two-dimensional phase space generated by the two fictitious magnetic fluxes. This local curvature is called Berry curvature after its discoverer, mathematical physicist Michael Berry. In particular, Simon showed that the Hall conductance was equal to $h/2\Pi$ times the local curvature at the origin of that phase space. This was a big deal. A famous mathematical result from 1848—the Gauss-Bonnet theorem—declared that the total curvature of a geometric shape was a topological feature, not a geometric one. In other words, the sum of all the local curvatures of a three-dimensional shape is the same for all topologically equivalent shapes with the same surface area. Even more exciting, the total curvature is simply given by $2\Pi(2 - 2g)$, where g is the number of holes in the shape.

Most important for us, a modern generalization of Gauss-Bonnet by geometer Shiing-shen Chern showed that the same result applied for the total Berry curvature of our two-dimensional phase space describing the quantum Hall effect. The Berry curvature of that space was now given by $2\Pi C$, with C denoting an integer known as the first Chern number. To show that the Hall conductance was quantized, Simon and his collaborators looked at the average of the conductance over the whole phase space, which is given by

$h/2\Pi$ times (total curvature) divided by (surface area). Plugging in $2\Pi C$ for the total curvature and $(h/e)^2$ for the surface area yielded $C \times e^2/h$. Et voilà. The average Hall conductance was an integer multiple of e^2/h, as Thouless had shown. But for the first time ever, the integer in front of e^2/h was identified with a "topological invariant"—a property that does not change if you rotate or deform a shape—and therefore the result was impervious to small perturbations and imperfections in the physical setup of the quantum Hall effect. This was a breakthrough insight.

Unfortunately, the beauty of the preceding arguments by Thouless and Simon was marred by a serious issue: the Hall conductance that experimentalists measured corresponded to the local curvature at the origin of the two-dimensional phase space, not the average curvature over the whole space. To see why the local curvature of an arbitrary shape is almost never equal to its average curvature, consider a torus. Gauss-Bonnet implies that the average curvature of a torus, and of any shape with a single hole in it, is zero. But the local curvature of a torus is obviously nonzero along most points on the surface and can take both positive and negative values. Thouless and his collaborators actually tried to address this issue, yet the question remained: Why was the Hall conductance quantized, if one was not allowed to average over all possible initial conditions of Laughlin's pump? Indeed, that was the question I had to answer.

A Sense of Despair

My first steps into the mystery of the quantum Hall effect were supposed to be illuminated by a book written by Thouless himself: *Topological Quantum Numbers in Nonrelativistic Physics*. A couple of weeks after receiving the book from Matt, I determined that I did not have the background required to understand any of the physics within. I locked the book inside my desk drawer and put the key away. Yet the book's simple existence gave me a sense of despair. How could I make any progress in solving the problem if I could not understand the contents of that book? Back then, I was a blank slate.

Of course, I had the option of going to Matt for help. He could teach me what I needed to know. Heck, we could even work closely on the problem together. But about a month or two after I arrived at Los Alamos, Matt told me he was leaving the lab. With job interviews now taking up most of his time, I barely saw him. A few months later, when he was offered a position at Microsoft's Station Q in Santa Barbara, Calif., my interactions with him all but ended. The few times we did meet, I became convinced that Matt had made a serious mistake in giving me a postdoc at Los Alamos. He would speak, and all I could retain were a few word combinations here and there. One of the phrases he repeated was "quasi-adiabatic continuation," a notion I was unfamiliar with. To my further dismay, this term did not seem to appear anywhere in the immense literature devoted to the quantum Hall effect up to that point.

Without much else to go on, I did what every young scientist of my generation would do and googled "quantum Hall effect" and "quasi-adiabatic continuation" (QAC). The first phrase returned hundreds of research papers, but I had as much luck reading through any of them as with the book by Thouless. The one thing I did get out of that search, however, was a word that kept coming up in relation to the quantum Hall effect: topological. When I added that word to my search, the first thing that popped up was an article by Avron, Osadchy and Seiler entitled "A Topological Look at the Quantum Hall Effect." The piece, which appeared in *Physics Today* in August 2003, was meant for nonexpert physicists. This article was so clearly written that it formed the foundation on which I would build my understanding of the quantum Hall effect.

In contrast to the hundreds of articles on the quantum Hall effect, my search on quasi-adiabatic continuation returned just two results, both by Matt. The first paper, co-authored with theoretical physicist Xiao-Gang Wen, was an introduction to QAC. The second paper contained, among other applications, a brief section on using QAC to compute a version of the Berry curvature relevant

Section 2: Time and Space

to the fractional quantum Hall effect. This was the first and only published attempt to apply QAC to any type of Berry curvature. I was excited to study Matt's argument inside and out. But I still needed to understand what QAC was about and how it was connected to adiabatic evolution. So I delved into the first paper, and after a month of poring over it, I felt that I had a good grasp of the technique. QAC was proposed as an evolution of a quantum system designed to preserve certain topological properties of its quantum state. In contrast, adiabatic evolution was better suited for local, geometric properties, such as the Berry curvature mentioned earlier.

The next task was to figure out how to compute the Berry curvature using QAC. To my dismay, I could not parse Matt's brief argument on how the two concepts could be bridged. I decided to re-create that bridge (or my version of it, at least) from scratch. The idea was to follow Simon's argument connecting adiabatic evolution to Berry curvature, while sneaking in QAC in place of adiabatic evolution. Substituting one evolution for the other worked out beautifully for one simple reason: I could show that QAC was exactly the same as adiabatic evolution under the following special condition: throughout the evolution of the system, the gap in energy between the ground state and the first excited state had to remain above a fixed positive value, independent of the size of the system. As luck would have it, this special condition was satisfied precisely near the origin of the 2-D phase space. In fact, if that condition was violated, I could show that the Hall conductance was not quantized.

After going through the exercise of connecting QAC to the Berry curvature and, hence, to the Hall conductance, I turned my sights toward the next big hurdle: re-creating Simon's argument, which computed the averaged Hall conductance as an unchanging topological quantity that yields the first Chern number. This was no small feat. As I have mentioned, to get over the initial problem of simulating adiabatic evolution with QAC, I took advantage of the fact that QAC tracked adiabatic evolution exactly, as long as there was a big enough spectral gap between the ground and excited states of the system. Unfortunately, this assumption about the spectral gap went out the window the moment I started exploring deeper into the 2-D phase space, whose total curvature I needed to compute. In fact, this assumption was so powerful that all attempts to quantize Hall conductance up to that point had used it. In other words, nobody thought it was possible to prove quantization without making that extra assumption. And neither did I. When I finally reached out to Matt in late spring of 2009 with a solution that made use of that key assumption, he said to me: "Nice job. But I think you should be able to prove quantization without it." Matt pointed me toward a seemingly unrelated paper of his entitled "Lieb-Schultz-Mattis in Higher Dimensions" (LSM), where he had laid the foundations for removing this assumption.

As I began to read through LSM, I had the same sinking feeling as when I had tried to parse Matt's attempt at connecting QAC to the Berry curvature. Deciphering it in isolation was going to be a long and arduous journey. But in a second twist of fate, my Ph.D. adviser, Bruno Nachtergaele, working with one of his postdocs at the time, Robert Sims, had published what some considered a mathematically rigorous version of Matt's LSM paper. Although most of the brilliant insights were already in Matt's original paper, Bruno's version was so well written and thorough that within a month I had a clear view of how to proceed. I now knew how to adapt elements of the LSM argument to overcome the second hurdle: to show that the averaged Hall conductance computed using QAC, instead of adiabatic evolution, was still an integer multiple of e^2/h.

The original Laughlin's pump argument, which used adiabatic evolution and gauge invariance to deduce a return to the original state of the system after one cycle, did not work with QAC. The main problem was that under QAC, after a flux quantum was inserted, there was no longer any guarantee that the system would end up in the same quantum state at the end of a cycle. Adiabatic evolution accomplished such a feat by forbidding the lowest-energy state of the system from ever getting excited. QAC, on the other hand, had a mind of its own. If the spectral gap ever dropped below a critical value as scientists inserted more and more magnetic flux, QAC would happily allow the system to jump to a new, excited quantum state, leaving behind its low-energy past. Unfortunately for me, that meant that at the end of a Laughlin cycle, even though the dynamics describing the system returned to their original state, the quantum state of the system itself may have changed significantly. If that were the case, then a key element of Laughlin's and Thouless's arguments would go up in smoke.

To overcome this obstacle, I needed to introduce two more fictitious magnetic fluxes in addition to the original two (for a grand total of four), which allowed me to transform the evolution under QAC into one that guaranteed a safe return to the original ground state at the end of a cycle. This trick, borrowed from Matt's LSM paper, forced the state of the system to maintain the exact same energy throughout the modified evolution around the boundary of the 2-D phase space, even when that energy no longer corresponded to the lowest possible energy of the system. In other words, to guarantee the return of the system to its initial state, all one ever needed to know was that the two states had the same energy. The fact that the ground state of the system was uniquely specified by that energy took care of the rest. Adiabatic evolution's insistence on keeping the system in its lowest-energy state throughout the evolution was overkill. More important, as I came to realize later, the insistence on using adiabatic evolution to quantize the Hall conductance was also the reason progress had stalled for nearly two decades.

By now I felt exhausted. But the main hurdle was finally in view. Everything I had accomplished up to this point was a fancy way of showing what Thouless, Simon and their collaborators had already proved: that the averaged Hall conductance was indeed quantized in integer multiples of e^2/h. It would seem that I had made no progress in removing the averaging assumption plaguing every effort to explain the mystery of the integer quantum Hall effect. Except for one minor detail: the two-dimensional phase space generated

Is There More Than One Universe?

by QAC had near-perfect uniform Berry curvature. In other words, the real Hall conductance, the one corresponding to the Berry curvature of a tiny patch near the origin of the 2-D phase space, was equal to the average curvature over the total flux space. Because the latter was famously quantized, it followed that the actual Hall conductance was also quantized. *Quod erat demonstrandum*—QED.

This final theoretical hurdle took many months of restless days and sleepless nights to cross over. I nearly gave up several times before reaching my goal. During a particularly dark time, I told my mom that I was not sure I wanted to wake up the next morning. In typical Greek fashion, she responded, "If you do anything stupid, I will fly out there and strangle you with my own two hands." Lost in a world of hyperanalytical thinking, I needed such an absurd statement to snap me out of it. I finished the proof in November 2009, shared it with Matt, who quickly added a section on how the result could be extended to also explain the fractional quantum Hall effect, and then posted it online. It would take us five more years before getting the result published and another four years before the mathematical physics community could fully digest it. On February 25, 2018, I opened an e-mail from Michael Aizenman—a letter I had waited for eight years to receive. It read:

Dear Matt and Spiros,
The Open Problems in Mathematical Physics web page was now updated with the statement that the IQHE question, which was posted by Yosi Avron and Ruedi Seiler, was solved in your joint work.

    I thank you here for your contribution, and also congratulate you on it. It is a pleasure to note that in each of the two problems on which progress is reported there, the advance came through deep novel insights and new tools. The list of solvers is a veritable honor roll.

Section 2: Time and Space

The fundamental mystery we started with was the question of why a microscopic, quantum phenomenon was showing up on a macroscopic scale. Instead what we found was that one of the most fundamental constants of nature was the reflection of global order beyond our finite grasp—the infinite communing with the infinitesimal. And although we have focused on the theory behind the quantum Hall effect, the experimental efforts it has inspired over the past three decades have been equally, if not more, exciting. Research on topological phases of matter beyond two-dimensional quantum Hall systems is paving the way toward technologies such as large-scale, fault-tolerant quantum computing. Impressive results coming out of labs such as Ana Maria Rey's at the University of Colorado Boulder are even tackling fundamental questions about the very nature of time.

This experience also taught me a valuable lesson: my self-worth is not tied to my success in life. The fateful call with my mom took place three months before I put the finishing touches on the solution. I did not turn into a mathematical genius within the span of a few months. But I made progress by breaking the problem down into simple parts I could understand. To do that, I needed to be okay with feeling incompetent most of the time. Without the faith of my parents in me as a person, whether I was good enough to solve the problem or not, I would have given up right before the finish line. Had I done that, the problem may still be unsolved and Marvel's Avengers would have had to find an even more scientifically implausible way to save the universe than to jump into the quantum realm via a macroscopic portal.

Referenced

Quantization of Hall Conductance for Interacting Electrons on a Torus. Matthew B. Hastings and Spyridon Michalakis in *Communications in Mathematical Physics*, Vol. 334, No. 1, pages 433–471; February 2015. arxiv.org/abs/0911.4706

Open Problems in Mathematical Physics: http:web.math.princeton.edu/~aizenman/OpenProblems_MathPhys/index.html

About the Author

Spyridon Michalakis is a mathematical physicist and manager of outreach for the Institute for Quantum Information and Matter at the California Institute of Technology.

Quantum Gravity in Flatland

By Steven Carlip

From its earliest days as a science, physics has searched for unity in nature. Isaac Newton showed that the same force responsible for the fall of an apple also holds the planets in their orbits. James Clerk Maxwell combined electricity, magnetism and light into a single theory of electromagnetism; a century later physicists added the weak nuclear force to form a unified "electroweak" theory. Albert Einstein joined space and time themselves into a single spacetime continuum.

Today the biggest missing link in this quest is the unification of gravity and quantum mechanics. Einstein's theory of gravity, his general theory of relativity, describes the birth of the universe, the orbits of planets and the fall of Newton's apple. Quantum mechanics describes atoms and molecules, electrons and quarks, the fundamental subatomic forces, and much besides. Yet in the places where both theories should apply—where both gravity and quantum effects are strong, such as black holes—they also seem incompatible.

Physicists' best efforts to create a single, unified theory that explains both quantum phenomena and gravity have failed miserably, giving answers that make no sense or no answers at all. Despite 80 years of work by physicists, including a dozen or so Nobel laureates, a quantum theory of gravity remains elusive.

Ask a physicist too hard a question, and a common reply will be, "Ask me something easier." Physics moves forward by looking at simple models that capture pieces of a complex reality. Researchers have worked on numerous such models for quantum gravity, including approximations that apply when gravity is weak or in special cases such as black holes.

Perhaps the most unusual approach is to neglect a whole dimension of space and work out how gravity would operate if our universe were only two-dimensional. (Technically, physicists refer to this situation as "(2+1)-dimensional," meaning two dimensions of space plus one of time.) The principles that govern gravity in this simplified universe might also apply to our 3-D one, thus giving us some much needed clues to unification.

The idea of dropping down a dimension has a distinguished history. Edwin Abbott's 1884 novel *Flatland: A Romance of Many Dimensions* follows the adventures of "A Square," a resident of a 2-D world of triangles, squares and other geometric figures. Although Abbott intended it as a satirical commentary on Victorian society—*Flatland* had a rigid class hierarchy, with linear women at the bottom and a class of circular priests at the top—*Flatland* also triggered a surge of interest in geometry in diverse dimensions and remains popular today among mathematicians and physicists.

Researchers trying to wrap their minds around a higher-dimensional realm start by imagining what our 3-D world would look like to A Square. *Flatland* has also inspired physicists studying materials such as graphene that really do behave like 2-D spaces.

The first studies of Flatland gravity, made in the early 1960s, were a letdown. A 2-D space literally would not have enough room for changes in the gravitational field to propagate. In the late 1980s, however, the subject had a renaissance as researchers realized that gravity works in unexpected ways. It would still sculpt the overall shape of space and even create black holes.

Flatland gravity has been a case study in lateral thinking, letting us subject some of our speculative ideas, such as the so-called holographic principle and the emergence of time from timelessness, to a rigorous mathematical test.

Time Management

When physicists seek to develop a quantum theory of a force, we take the corresponding classical theory as our starting point and build on it. For gravity, that means general relativity, and there the trouble starts. General relativity involves a complex system of 10 equations, each with up to thousands of terms. We cannot solve these equations in their full generality, so we face a daunting task in formulating their quantum version. But the mystery of why quantum gravity is so elusive is deeper still.

According to general relativity, the thing we call "gravity" is actually a manifestation of the shape of space and time. Earth orbits the sun not because some force tugs on it but because it is moving along the straightest possible path in a spacetime that has been warped by the sun's mass. Uniting quantum mechanics and gravity means somehow quantizing the structure of space and time itself.

That may not sound so challenging. Yet a cornerstone of quantum mechanics is the Heisenberg uncertainty principle, the idea that physical quantities are inherently fuzzy—fluctuating randomly and having no definite values unless they are observed or undergo an equivalent process. In a quantum theory of gravity, space and time themselves fluctuate, shaking the scaffolding on which the rest of physics is built. Without a fixed spacetime as the background, we do not know how to describe positions, rates of change or any of the other basic quantities of physics. Simply put, we do not know what a quantum spacetime means.

These general obstacles to conceptualizing quantized spacetime show up in several specific ways. One is the notorious "problem of time." Time is fundamental to our observed reality. Almost every theory of physics is ultimately a description of the way some piece of the universe changes in time. So we physicists had better know what "time" means, and the embarrassing truth is that we do not.

To Newton, time was absolute—standing outside nature, affecting matter but unaffected by it. The usual formulations of quantum mechanics accept this idea of an absolute time. Relativity, however, dethroned absolute time. Different observers in relative motion disagree about the passage of time and even about whether two events are simultaneous. A clock—as well as anything else that varies in time—runs more slowly in a strong gravitational field. No longer merely an external parameter, time is now an active participant in the universe. But if there is no ideal clock sitting outside the universe and determining the pace of change, the passage of time must arise from the internal structure of the universe. But how? It is hard to even know where to start.

The problem of time has a less famous cousin, the problem of observables. Physics is an empirical science; a theory must make verifiable predictions for observable quantities. In ordinary physics, these quantities are ascribed to specific locations: the strength of the electric field "here" or the probability of finding an electron "there." We label "here" or "there" with the coordinates x, y and z, and our theories predict how observables depend on the values of these coordinates.

Yet according to Einstein, spatial coordinates are arbitrary, human-made labels, and in the end the universe does not care about them. If you cannot identify a point in spacetime objectively, then you cannot claim to know what is going on at it. Charles Torre of Utah State University has shown that a quantum theory of gravity can have no purely local observables—that is, observables whose values depend on only a single point in spacetime. So scientists are left with nonlocal observables, quantities whose values depend on many points at once. In general, we do not even know how to define such objects, much less use them to describe the world we observe.

A third problem is how the universe came into being. Did it pop into existence from nothing? Did it split off a parent universe? Or did it do something else entirely? Each possibility poses some difficulty for a quantum theory of gravity. A related problem is a perennial favorite of science-fiction writers: wormholes, which form shortcuts between locations in space or even in time. Physicists have thought seriously about this idea—in the past 20 years they have written more than 1,000 journal articles on wormholes—without settling the question of whether such structures are possible.

A final set of questions revolves around the most mysterious beasts known to science: black holes. They may offer our best window into the ultimate nature of space and time. In the early 1970s Stephen Hawking showed that black holes should glow like a hot coal—emitting radiation with a so-called blackbody spectrum. In every other physical system, temperature reflects the underlying behavior of microscopic constituents. When we say a room is hot, what we really mean is that the molecules of air inside it are moving energetically. For a black hole the "molecules" must be quantum-gravitational. They are not literally molecules but some unknown microscopic substructure—what a physicist would call "degrees of freedom"—that must be capable of changing. No one knows what they truly are.

An Unattractive Model

At first glance, Flatland seems an unpromising place to seek answers to these questions. Abbott's Flatland had many laws, but a law of gravity was not among them. In 1963 Polish physicist Andrzej Staruszkiewicz worked out what that law might be by applying general relativity. He found that a massive object in Flatland would bend the 2-D space around it into a cone, like a party hat made by twisting a flat piece of paper. A small object passing the apex of this cone would find its path deflected, much as the sun bends a comet's path in our universe. In 1984 Stanley Deser of Brandeis University, Roman Jackiw of the Massachusetts Institute of Technology and Gerard 't Hooft of Utrecht University in the Netherlands worked out how quantum particles would move through such a space.

This geometry would be much simpler than the complicated pattern of curvature that gravity causes in our 4-D spacetime. Flatland would lack the equivalent of Newton's law of attraction; instead the strength of the force would depend on objects' velocities, and two bodies at rest would not be pulled toward each other. This simplicity is appealing. It suggests that quantizing Staruszkiewicz's theory would be easier than quantizing full-blown general relativity in 3-D. Unfortunately, the theory is too simple: nothing is left to quantize. A 2-D space has no room for an important element of Einstein's theory: gravitational waves.

Consider the simpler case of electromagnetism. Electric and magnetic fields are produced by electric charges and currents. As Maxwell showed, these fields can detach themselves from their sources and move freely as light waves. In the quantum version of Maxwell's theory, the waves become photons, the quanta of light. In the same way, the gravitational fields of general relativity can detach themselves from their sources and become freely propagating gravitational waves, and physicists widely assume that a quantum theory of gravity will contain particles called gravitons that do the traveling.

A light wave has a polarization: its electric field oscillates in a direction perpendicular to its direction of motion. A gravitational wave also has a polarization, but the pattern is more complicated: the field oscillates not in one but in two directions perpendicular to its direction of motion. Flatland has no room for this behavior. Once the direction of motion is fixed, only one perpendicular direction remains. Gravitational waves and their quantum counterparts, gravitons, simply cannot be squeezed into just two dimensions of space.

Despite occasional sparks of interest, Staruszkiewicz's discovery languished. Then, in 1989, Edward Witten of the Institute for Advanced Study in Princeton, N.J., stepped in. Witten, widely considered the world's leading mathematical physicist, had been working on a special class of fields in which waves do not propagate freely. When he realized that 2-D gravity fit into this class, he added the crucial missing ingredient: topology.

Doughnutland

Witten pointed out that even if gravity cannot propagate as waves, it can still have a dramatic effect on the overall shape of space. This effect does not arise when Flatland is just a plane; it requires a more complex topology. When an ice sculpture melts away, the

Section 2: Time and Space

details become muted, but certain features such as holes tend to last. Topology describes these features.

Two surfaces have the same topology if one can be smoothly deformed into the other without cutting, tearing or gluing. For instance, a hemisphere and a disk share the same topology: stretching the hemisphere by pulling on its perimeter yields a disk. A sphere has another topology: to turn it into a hemisphere or disk, you would need to snip out a piece. A torus, like the surface of a doughnut, has yet another. The surface of a coffee cup has the same topology as a torus: the handle looks like a torus, and the rest of the cup can be smoothed out without cutting or tearing—hence the old mathematician's joke that a topologist can't tell a doughnut from a coffee cup.

Although tori look curved, when you consider their internal geometry rather than their shape as seen from the outside, they can actually be flat. What makes a torus a torus is the fact you can make a full loop around it in two separate directions: through the hole or around the rim. This feature will be familiar to anyone who has played any 1980s-era video game in which a combatant exiting the right side of the screen reenters on the left. The screen is flat: it obeys the rule of plane geometry, such as the fact that parallel lines never meet. Yet the topology is toroidal.

In fact, an infinite family of such tori exist—all flat but all distinct, labeled by a parameter called the modulus. What gravity in a toroidal universe does is to cause the modulus to evolve in time. A torus starts as a line at the big bang and opens up to assume an ever more square-shaped geometry as the universe expands.

Starting with Witten's results, I showed that this process could be quantized—and that doing so turns the classical theory of gravity into a quantum one. Quantum gravity in Flatland is a theory not of gravitons but of shape-shifting tori. That view is a shift from the usual picture of quantum theory as a theory of the very small. Quantum gravity in two dimensions is, in fact, a theory of the entire universe as a single object. This insight gives us a rich enough model to explore some of the fundamental conceptual problems of quantum gravity.

Finding the Time

Flatland gravity demonstrates, for example, how time might emerge from a fundamentally timeless reality. In one formulation of the theory, the entire universe is described by a single, quantum, wave function, similar to the mathematical device that physicists routinely use to describe atoms and subatomic particles. This wave function does not depend on time, because it already includes all time—past, present and future—in one package. Somehow this "timeless" wave function gives rise to the change we observe in the world. The trick is to remember Einstein's aphorism that time is what is measured by a clock. Time does not stand outside the universe; it is determined by a subsystem that is correlated with the rest of the universe, just as a wall clock is correlated with Earth's rotation.

The theory offers many different clock options, and our choice defines what we mean by "time." In Doughnutland, A Square can define time by using the readings of atomic clocks in satellites, like those in the GPS. He can label time by the lengths of curves extending from the big bang, by the size of his expanding universe, or by the amount of redshift caused by its expansion. Once he has made such a choice, all other physical observables change with clock time. The modulus of the torus universe is correlated with its size, for instance, and A Square perceives this as a universe evolving in time. The theory thus bootstraps time from a timeless universe. These ideas are not new, but quantum gravity in Doughnutland has at last given us a setting in which we can do the math and check that the picture does not just look pretty but really works. Some of the definitions of time have intriguing consequences, such as implying that space can be creased.

As for the problem of observables, Doughnutland gives us a set of objectively measurable quantities—namely, the moduli. The twist is that these quantities are nonlocal: they do not reside at specific locations but describe the structure of the whole space. Anything that A Square measures is ultimately a proxy for these nonlocal quantities. In 2008 Catherine Meusburger, now at the University of Erlangen-Nürnberg in Germany, showed how these moduli relate to real cosmological measurements such as time delays and redshifts for beams of light. I have shown how they relate to objects' motion.

Flatland gravity offers good news for fans of wormholes: at least one formulation of the theory permits the topology of space to change. A Square could go to bed tonight in Sphereland and wake up tomorrow in Doughnutland, which is equivalent to creating a shortcut between two distant corners of the universe. In some versions of the theory, we can describe the creation of the universe out of nothing, the ultimate change in topology.

On the Edge of Space

Because gravity in flatland is stunted, it used to be common knowledge among experts in the field (me included) that 2-D black holes were impossible. But in 1992 three physicists—Máximo Bañados, now at the Pontifical Catholic University of Chile in Santiago, and Claudio Bunster (then Claudio Teitelboim) and Jorge Zanelli, both at the Center for Scientific Studies in Valdivia, Chile—shocked the world, or at least our little corner of it, by showing that the theory does allow black holes, as long as the universe has a certain type of dark energy.

A so-called BTZ black hole is very much like a real black hole in our own universe. Formed from matter collapsing under its own weight, it is surrounded by an event horizon, a one-way barrier from which nothing can escape. To an observer who remains on the outside, the event horizon looks like an edge of the universe: any object that falls through the horizon is completely cut off from us. Per Hawking's calculations, A Square should see it glow at a temperature that depends on its mass and spin.

That result presents a puzzle. Lacking gravitational waves or gravitons, Flatland gravity should also lack the gravitational degrees of freedom that would explain black hole temperature. Yet they sneak in anyway. The reason is that the event horizon itself provides some additional structure that empty 2-D space lacks. The horizon exists at a certain location, which, mathematically, augments the raw theory with some additional quantities. Vibrations that wiggle the horizon provide degrees of freedom. Remarkably, we find that they exactly reproduce Hawking's results.

Because the degrees of freedom are features of the horizon, they reside, in a sense, on the edge of Flatland itself. So they are a concrete realization of a fascinating proposal about the nature of quantum gravity, the holographic principle. This principle suggests that dimension may be a fungible concept. Just as a hologram captures a three-dimensional image on a flat 2-D film, many physicists speculate that the physics of a d-dimensional world can be completely captured by a simpler theory in d–1 dimensions. In string theory—a leading effort to unify general relativity and quantum mechanics—this idea led in the late 1990s to a novel approach for creating a quantum theory of gravity.

Flatland gravity provides a simplified scenario to test that approach. In 2007 Witten and Alexander Maloney, now at McGill University, again surprised the physics world by suggesting that the holographic predictions appear to fail for the simplest form of 2-D gravity. The theory, they found, seemed to predict impossible thermal properties for black holes. This unexpected result suggests that gravity is an even more subtle phenomenon than we had suspected, and the response has been a fresh surge of Flatland research.

Perhaps gravity simply does not make sense by itself but must work in partnership with other kinds of forces and particles. Perhaps Einstein's theory needs to be revised. Perhaps we need to find a way to put back some local degrees of freedom. Perhaps the holographic principle does not always hold. Or perhaps space, like time, is not a fundamental ingredient of the universe. Whatever the answer, Flatland gravity has pointed us in a direction we might not otherwise have taken.

Although we cannot make a real 2-D black hole, we might be able to test some of the predictions of the Flatland model experimentally.

Several laboratories around the world are working on 2-D analogues of black holes. For example, a fluid flowing faster than the speed of sound produces a sonic event horizon from which sound waves cannot escape. Experimenters have also built 2-D black holes by using electromagnetic waves that are confined to surfaces. Such analogues should also exhibit a quantum glow in much the same way a black hole does.

Quantum gravity in Flatland began as a playground for physicists, a simple setting in which to explore ideas about real-world quantum gravity. It has already taught us valuable lessons about time, observables and topology that are carrying over to real 3-D gravity. The model has surprised us with its richness: the unexpectedly important role of topology, its remarkable black holes, its strange holographic properties. Perhaps soon we will fully understand what it is like to be a square living in a flat world.

Referenced

Quantum Gravity in 2+1 Dimensions. Steven Carlip. Cambridge University Press, 1998.

The Planiverse: Computer Contact with a Two-Dimensional World. A. K. Dewdney. Springer, 2001.

Quantum Gravity in 2+1 Dimensions: The Case of a Closed Universe. Steven Carlip in *Living Reviews in Relativity*, Vol. 8; 2005. www.livingreviews.org/lrr-2005-1

Steven Carlip's Web page offers a glossary of quantum gravity: www.physics.ucdavis.edu/Text/Carlip.html#Glos

About the Author

Steven Carlip worked as a printer, a newspaper editor and a factory worker before deciding to become a physicist. He studied under Bryce DeWitt, one of the founding figures of quantum gravity, and is now a professor at the University of California, Davis. He is a fellow of the American Physical Society and of its British counterpart, the Institute of Physics.

The Beauty at the Heart of a "Spooky" Mystery

By John Horgan

Studying quantum mechanics, which I've been doing for the last two-plus years, has served as an antidote to my tendency toward habituation, taking reality for granted. Wave functions, superposition and other esoterica remind me that this is a strange, strange world; there is a mystery at the heart of things that ordinary language can never quite capture.

I'm thus thrilled by this year's Nobel Prize for Physics. John Clauser, Alain Aspect and Anton Zeilinger won for experimental probes of entanglement, a peculiar connection between two or more particles. The Nobel Foundation's press release emphasizes the applications of this prize-winning work; researchers are building "quantum computers, quantum networks and secure quantum encrypted communication" based on entanglement. But I value the work of Clauser, et al., because it upends our commonsense notions about what is real and what is knowable. It rubs our noses in the riddle of reality.

Experts bicker over what entanglement is and what it means; philosopher of physics Tim Maudlin complains that the Nobel Committee for Physics misunderstands entanglement. My "understanding," such as it is, begins with wave functions, mathematical widgets that describe the behavior of electrons, photons and other quantum stuff. Unlike, say, Newton's laws of motion, which precisely track objects' trajectories, a wave function tracks only the *probability* that an electron, say, will behave in a certain way. The probabilities undulate over time in wavelike fashion; hence the term.

When you look at the electron—measuring it with some sort of instrument—its wave function is said to *collapse*, and you see only one of the possible outcomes. That is strange enough. Even stranger

Section 2: Time and Space

is what happens when the wave function applies to two or more particles that start out conjoined in a particular way. Imagine you have a wave function describing a radioactive lump that emits two electrons at the same time. Call the electrons A and B.

Electrons possess a quantum property called spin, which is unlike the spin of a planet or top. Quantum spin is binary; it is either up or down, to use a common notation. Imagine if planets could only spin clockwise, or counterclockwise, with their axes pointed only at the North Star, and in no other direction, and you're getting the gist of spin. Although quantum spin, like entanglement, makes no sense, it has been verified countless times over the past century.

Okay, now you let the electrons fly apart from each other. Then you measure the spin of electron A and find that its spin is up. At that moment, the wave function for *both* electrons collapses, *instantaneously* predicting the spin of electron B, even if it is a light-year away. How can that be? How can your measurement of A tell you something about B instantaneously? Entanglement seems to violate special relativity, which says that effects cannot propagate faster than the speed of light. Entanglement also implies that the two electrons, before you measure them, do not have a fixed spin; they exist in a probabilistic limbo.

Einstein objected to entanglement, which he famously derided as "spooky action at a distance." Einstein felt that physics theories should possess two properties, sometimes called locality and realism. Locality says effects cannot propagate faster than the speed of light; realism says physical things, such as electrons, possess specific properties, such as spin or position, all the time and not just when we measure them. Einstein argued that if quantum mechanics violates realism and locality, it must be flawed, or incomplete.

For decades, the debate over entanglement was seen as purely philosophical, that is, experimentally unresolvable. Then in 1964, John Bell presented a mathematical argument that turned philosophy into physics. If your model of entanglement is based on locality and realism, Bell showed, it will produce results that differ, statistically, from those of quantum mechanics. This difference is called Bell's inequality.

John Clauser, Alain Aspect and Anton Zeilinger put Bell's theorem to the test, performing experiments on entangled photons and other particles. Their research has confirmed that the predictions of quantum mechanics hold up. The experiments dash the hopes of Einstein and others that causes and effects propagate in an orderly fashion, and that things have specific properties when we don't look at them.

John Bell died in 1990, too early to see his ideas fully vindicated—or to share the Nobel Prize, which is not given posthumously. But he left behind a collection of influential papers, collected under the title *Speakable and Unspeakable in Quantum Mechanics*. Ironically, quantum theorists cite Bell's utterances like scripture, even though his own views seem fluid, unsettled, riddled with self-doubt. He even disses his own inequality theorem, suggesting that "what is proved by impossibility proofs is lack of imagination." Bell's theorem is an impossibility proof.

Bell seems less intent on solving the paradoxes of quantum mechanics than on drawing attention to them. In a 1986 essay, he compares his fellow physicists to "sleepwalkers," who continue to extend quantum theory while ignoring its "fundamental obscurity." Given the "immensely impressive" progress achieved by sleepwalking physicists, Bell asks, "is it wise to shout, 'wake up'? I am not sure that it is. So I speak now in a very low voice."

Bell once said that quantum mechanics "carries in itself the seeds of its own destruction." He, like Einstein, seemed to hope that quantum mechanics would yield to a more sensible theory, ideally one that restores locality, realism and certainty to physics. My guess is that if we find such a theory, it will eventually turn out to be mysterious in its own way. The mystery might be unlike our quantum mystery, but it will still be a mystery, which cuts through our habituation and forces us to pay attention to the weird, weird world.

Section 2: Time and Space

This is an opinion and analysis article, and the views expressed by the author or authors are not necessarily those of Scientific American.

About the Author

John Horgan directs the Center for Science Writings at the Stevens Institute of Technology. His books include The End of Science, The End of War *and* Mind-Body Problems, *available for free at mindbodyproblems.com. For many years he wrote the popular blog* Cross Check *for* Scientific American.

This Twist on Schrödinger's Cat Paradox Has Major Implications for Quantum Theory

By Zeeya Merali

That question irked and inspired Hungarian-American physicist Eugene Wigner in the 1960s. He was frustrated by the paradoxes arising from the vagaries of quantum mechanics—the theory governing the microscopic realm that suggests, among many other counterintuitive things, that until a quantum system is observed, it does not necessarily have definite properties. Take his fellow physicist Erwin Schrödinger's famous thought experiment in which a cat is trapped in a box with poison that will be released if a radioactive atom decays. Radioactivity is a quantum process, so before the box is opened, the story goes, the atom has both decayed and not decayed, leaving the unfortunate cat in limbo—a so-called superposition between life and death. But does the cat experience being in superposition?

Wigner sharpened the paradox by imagining a (human) friend of his shut in a lab, measuring a quantum system. He argued it was absurd to say his friend exists in a superposition of having seen and not seen a decay unless and until Wigner opens the lab door. "The 'Wigner's friend' thought experiment shows that things can become very weird if the observer is also observed," says Nora Tischler, a quantum physicist at Griffith University in Brisbane, Australia.

Now Tischler and her colleagues have carried out a version of the Wigner's friend test. By combining the classic thought experiment with another quantum head-scratcher called entanglement—a phenomenon that links particles across vast distances—they have also derived a new theorem, which they claim puts the strongest constraints yet on the fundamental nature of reality. Their study, which appeared in *Nature Physics* on August 17, 2020, has implications for the role that consciousness might play in quantum physics—and even whether quantum theory must be replaced.

The new work is an "important step forward in the field of experimental metaphysics," says quantum physicist Aephraim Steinberg of the University of Toronto, who was not involved in the study. "It's the beginning of what I expect will be a huge program of research."

A Matter of Taste

Until quantum physics came along in the 1920s, physicists expected their theories to be deterministic, generating predictions for the outcome of experiments with certainty. But quantum theory appears to be inherently probabilistic. The textbook version—sometimes called the Copenhagen interpretation—says that until a system's properties are measured, they can encompass myriad values. This superposition only collapses into a single state when the system is observed, and physicists can never precisely predict what that state will be. Wigner held the then popular view that consciousness somehow triggers a superposition to collapse. Thus, his hypothetical friend would discern a definite outcome when she or he made a measurement—and Wigner would never see her or him in superposition.

This view has since fallen out of favor. "People in the foundations of quantum mechanics rapidly dismiss Wigner's view as spooky and ill-defined because it makes observers special," says David Chalmers, a philosopher and cognitive scientist at New York University. Today most physicists concur that inanimate objects can knock quantum systems out of superposition through a process known as decoherence. Certainly, researchers attempting to manipulate complex quantum superpositions in the lab can find their hard work destroyed by speedy air particles colliding with their systems. So they carry out their tests at ultracold temperatures and try to isolate their apparatuses from vibrations.

Several competing quantum interpretations have sprung up over the decades that employ less mystical mechanisms, such as

Is There More Than One Universe?

decoherence, to explain how superpositions break down without invoking consciousness. Other interpretations hold the even more radical position that there is no collapse at all. Each has its own weird and wonderful take on Wigner's test. The most exotic is the "many worlds" view, which says that whenever you make a quantum measurement, reality fractures, creating parallel universes to accommodate every possible outcome. Thus, Wigner's friend would split into two copies and, "with good enough supertechnology," he could indeed measure that person to be in superposition from outside the lab, says quantum physicist and many-worlds fan Lev Vaidman of Tel Aviv University.

The alternative "Bohmian" theory (named for physicist David Bohm) says that at the fundamental level, quantum systems do have definite properties; we just do not know enough about those systems to precisely predict their behavior. In that case, the friend has a single experience, but Wigner may still measure that individual to be in a superposition because of his own ignorance. In contrast, a relative newcomer on the block called the QBism interpretation embraces the probabilistic element of quantum theory wholeheartedly (QBism, pronounced "cubism," is actually short for quantum Bayesianism, a reference to 18th-century mathematician Thomas Bayes's work on probability.) QBists argue that a person can only use quantum mechanics to calculate how to calibrate his or her beliefs about what he or she will measure in an experiment. "Measurement outcomes must be regarded as personal to the agent who makes the measurement," says Ruediger Schack of Royal Holloway, University of London, who is one of QBism's founders. According to QBism's tenets, quantum theory cannot tell you anything about the underlying state of reality, nor can Wigner use it to speculate on his friend's experiences.

Another intriguing interpretation, called retrocausality, allows events in the future to influence the past. "In a retrocausal account, Wigner's friend absolutely does experience something," says Ken Wharton, a physicist at San Jose State University, who is an advocate for this time-twisting view. But that "something" the friend experiences at the point of measurement can depend upon Wigner's choice of how to observe that person later.

The trouble is that each interpretation is equally good—or bad—at predicting the outcome of quantum tests, so choosing between them comes down to taste. "No one knows what the solution is," Steinberg says. "We don't even know if the list of potential solutions we have is exhaustive."

Other models, called collapse theories, do make testable predictions. These models tack on a mechanism that forces a quantum system to collapse when it gets too big—explaining why cats, people and other macroscopic objects cannot be in superposition. Experiments are underway to hunt for signatures of such collapses, but as yet they have not found anything. Quantum physicists are also placing ever larger objects into superposition: last year a team in Vienna reported doing so with a 2,000-atom molecule. Most quantum interpretations say there is no reason why these efforts to supersize superpositions should not continue upward forever, presuming researchers can devise the right experiments in pristine lab conditions so that decoherence can be avoided. Collapse theories, however, posit that a limit will one day be reached, regardless of how carefully experiments are prepared. "If you try and manipulate a classical observer—a human, say—and treat it as a quantum system, it would immediately collapse," says Angelo Bassi, a quantum physicist and proponent of collapse theories at the University of Trieste in Italy.

A Way to Watch Wigner's Friend

Tischler and her colleagues believed that analyzing and performing a Wigner's friend experiment could shed light on the limits of quantum theory. They were inspired by a new wave of theoretical and experimental papers that have investigated the role of the observer in quantum theory by bringing entanglement into Wigner's classic setup. Say you take two particles of light, or photons, that are polarized so that they can vibrate horizontally or vertically. The

Is There More Than One Universe?

photons can also be placed in a superposition of vibrating both horizontally and vertically at the same time, just as Schrödinger's paradoxical cat can be both alive and dead before it is observed.

Such pairs of photons can be prepared together—entangled—so that their polarizations are always found to be in the opposite direction when observed. That may not seem strange—unless you remember that these properties are not fixed until they are measured. Even if one photon is given to a physicist called Alice in Australia, while the other is transported to her colleague Bob in a lab in Vienna, entanglement ensures that as soon as Alice observes her photon and, for instance, finds its polarization to be horizontal, the polarization of Bob's photon instantly syncs to vibrating vertically. Because the two photons appear to communicate faster than the speed of light—something prohibited by his theories of relativity—this phenomenon deeply troubled Albert Einstein, who dubbed it "spooky action at a distance."

These concerns remained theoretical until the 1960s, when physicist John Bell devised a way to test if reality is truly spooky—or if there could be a more mundane explanation behind the correlations between entangled partners. Bell imagined a commonsense theory that was local—that is, one in which influences could not travel between particles instantly. It was also deterministic rather than inherently probabilistic, so experimental results could, in principle, be predicted with certainty, if only physicists understood more about the system's hidden properties. And it was realistic, which, to a quantum theorist, means that systems would have these definite properties even if nobody looked at them. Then Bell calculated the maximum level of correlations between a series of entangled particles that such a local, deterministic and realistic theory could support. If that threshold was violated in an experiment, then one of the assumptions behind the theory must be false.

Such "Bell tests" have since been carried out, with a series of watertight versions performed in 2015, and they have confirmed reality's spookiness. "Quantum foundations is a field that was really started experimentally by Bell's [theorem]—now over 50 years old.

Section 2: Time and Space

And we've spent a lot of time reimplementing those experiments and discussing what they mean," Steinberg says. "It's very rare that people are able to come up with a new test that moves beyond Bell."

The Brisbane team's aim was to derive and test a new theorem that would do just that, providing even stricter constraints—"local friendliness" bounds—on the nature of reality. Like Bell's theory, the researchers' imaginary one is local. They also explicitly ban "superdeterminism"—that is, they insist that experimenters are free to choose what to measure without being influenced by events in the future or the distant past. (Bell implicitly assumed that experimenters can make free choices, too.) Finally, the team prescribes that when an observer makes a measurement, the outcome is a real, single event in the world—it is not relative to anyone or anything.

Testing local friendliness requires a cunning setup involving two "superobservers," Alice and Bob (who play the role of Wigner), watching their friends Charlie and Debbie. Alice and Bob each have their own interferometer—an apparatus used to manipulate beams of photons. Before being measured, the photons' polarizations are in a superposition of being both horizontal and vertical. Pairs of entangled photons are prepared such that if the polarization of one is measured to be horizontal, the polarization of its partner should immediately flip to be vertical. One photon from each entangled pair is sent into Alice's interferometer, and its partner is sent to Bob's. Charlie and Debbie are not actually human friends in this test. Rather, they are beam displacers at the front of each interferometer. When Alice's photon hits the displacer, its polarization is effectively measured, and it swerves either left or right, depending on the direction of the polarization it snaps into. This action plays the role of Alice's friend Charlie "measuring" the polarization. (Debbie similarly resides in Bob's interferometer.)

Alice then has to make a choice: She can measure the photon's new deviated path immediately, which would be the equivalent of opening the lab door and asking Charlie what he saw. Or she can allow the photon to continue on its journey, passing through a second beam displacer that recombines the left and right paths—the equivalent of keeping the lab door closed. Alice can then directly measure her photon's polarization as it exits the interferometer. Throughout the experiment, Alice and Bob independently choose which measurement choices to make and then compare notes to calculate the correlations seen across a series of entangled pairs.

Tischler and her colleagues carried out 90,000 runs of the experiment. As expected, the correlations violated Bell's original bounds—and crucially, they also violated the new local-friendliness threshold. The team could also modify the setup to tune down the degree of entanglement between the photons by sending one of the pair on a detour before it entered its interferometer, gently perturbing the perfect harmony between the partners. When the researchers ran the experiment with this slightly lower level of entanglement, they found a point where the correlations still violated Bell's bound but not local friendliness. This result proved that the two sets of bounds are not equivalent and that the new local-friendliness constraints are stronger, Tischler says. "If you violate them, you learn more about reality," she adds. Namely, if your theory says that "friends" can be treated as quantum systems, then you must either give up locality, accept that measurements do not have a single result that observers must agree on or allow superdeterminism. Each of these options has profound—and, to some physicists, distinctly distasteful—implications.

Reconsidering Reality

"The paper is an important philosophical study," says Michele Reilly, co-founder of Turing, a quantum-computing company based in New York City, who was not involved in the work. She notes that physicists studying quantum foundations have often struggled to come up with a feasible test to back up their big ideas. "I am thrilled to see an experiment behind philosophical studies," Reilly says. Steinberg calls the experiment "extremely elegant" and praises the team for tackling the mystery of the observer's role in measurement head-on.

Section 2: Time and Space

Although it is no surprise that quantum mechanics forces us to give up a commonsense assumption—physicists knew that from Bell—"the advance here is that we are a narrowing in on which of those assumptions it is," says Wharton, who was also not part of the study. Still, he notes, proponents of most quantum interpretations will not lose any sleep. Fans of retrocausality, such as himself, have already made peace with superdeterminism: in their view, it is not shocking that future measurements affect past results. Meanwhile QBists and many-worlds adherents long ago threw out the requirement that quantum mechanics prescribes a single outcome that every observer must agree on.

And both Bohmian mechanics and spontaneous collapse models already happily ditched locality in response to Bell. Furthermore, collapse models say that a real macroscopic friend cannot be manipulated as a quantum system in the first place.

Vaidman, who was also not involved in the new work, is less enthused by it, however, and criticizes the identification of Wigner's friend with a photon. The methods used in the paper "are ridiculous; the friend has to be macroscopic," he says. Philosopher of physics Tim Maudlin of New York University, who was not part of the study, agrees. "Nobody thinks a photon is an observer, unless you are a panpsychic," he says. Because no physicist questions whether a photon can be put into superposition, Maudlin feels the experiment lacks bite. "It rules something out—just something that nobody ever proposed," he says.

Tischler accepts the criticism. "We don't want to overclaim what we have done," she says. The key for future experiments will be scaling up the size of the "friend," adds team member Howard Wiseman, a physicist at Griffith University. The most dramatic result, he says, would involve using an artificial intelligence, embodied on a quantum computer, as the friend. Some philosophers have mused that such a machine could have humanlike experiences, a position known as the strong AI hypothesis, Wiseman notes, though nobody yet knows whether that idea will turn out to be true. But if the hypothesis holds, this quantum-based artificial general intelligence (AGI) would be microscopic. So from the point of view of spontaneous collapse models, it would not trigger collapse because of its size. If such a test was run, and the local-friendliness bound was not violated, that result would imply that an AGI's consciousness cannot be put into superposition. In turn, that conclusion would suggest that Wigner was right that consciousness causes collapse. "I don't think I will live to see an experiment like this," Wiseman says. "But that would be revolutionary."

Reilly, however, warns that physicists hoping that future AGI will help them home in on the fundamental description of reality are putting the cart before the horse. "It's not inconceivable to me that quantum computers will be the paradigm shift to get to us into AGI," she says. "Ultimately, we need a theory of everything in order to build an AGI on a quantum computer, period, full stop."

That requirement may rule out more grandiose plans. But the team also suggests more modest intermediate tests involving machine-learning systems as friends, which appeals to Steinberg. That approach is "interesting and provocative," he says. "It's becoming conceivable that larger- and larger-scale computational devices could, in fact, be measured in a quantum way."

Renato Renner, a quantum physicist at the Swiss Federal Institute of Technology Zurich (ETH Zurich), makes an even stronger claim: regardless of whether future experiments can be carried out, he says, the new theorem tells us that quantum mechanics needs to be replaced. In 2018 Renner and his colleague Daniela Frauchiger, then at ETH Zurich, published a thought experiment based on Wigner's friend and used it to derive a new paradox. Their setup differs from that of the Brisbane team but also involves four observers whose measurements can become entangled. Renner and Frauchiger calculated that if the observers apply quantum laws to one another, they can end up inferring different results in the same experiment.

"The new paper is another confirmation that we have a problem with current quantum theory," says Renner, who was not involved in the work. He argues that none of today's quantum interpretations can worm their way out of the so-called Frauchiger-Renner paradox without proponents admitting they do not care whether quantum theory gives consistent results. QBists offer the most palatable means of escape, because from the outset, they say that quantum theory cannot be used to infer what other observers will measure, Renner says. "It still worries me, though: If everything is just personal to me, how can I say anything relevant to you?" he adds. Renner is now working on a new theory that provides a set of mathematical rules that would allow one observer to work out what another should see in a quantum experiment.

Still, those who strongly believe their favorite interpretation is right see little value in Tischler's study. "If you think quantum mechanics is unhealthy, and it needs replacing, then this is useful because it tells you new constraints," Vaidman says. "But I don't agree that this is the case—many worlds explains everything."

For now, physicists will have to continue to agree to disagree about which interpretation is best or if an entirely new theory is needed. "That's where we left off in the early 20th century—we're genuinely confused about this," Reilly says. "But these studies are exactly the right thing to do to think through it."

Disclaimer: The author frequently writes for the Foundational Questions Institute, which sponsors research in physics and cosmology and partially funded the Brisbane team's study.

About the Author

Zeeya Merali is a freelance writer based in London and author of A Big Bang in a Little Room.

Can Quantum Mechanics Save the Cosmic Multiverse?

By Yasunori Nomura

Many cosmologists now accept the extraordinary idea that what seems to be the entire universe may actually be only a tiny part of a much larger structure called the multiverse. In this picture, multiple universes exist, and the rules we once assumed were basic laws of nature take different forms in each; for example, the types and properties of elementary particles may differ from one universe to another. The multiverse idea emerges from a theory that suggests the very early cosmos expanded exponentially. During this period of "inflation," some regions would have halted their rapid expansion sooner than others, forming what are called bubble universes, much like bubbles in boiling water. Our universe would be just one of these bubbles, and beyond it would lie infinitely more.

The idea that our entire universe is only a part of a much larger structure is, by itself, not as outlandish as it sounds. Throughout history scientists have learned many times over that the visible world is far from all there is. Yet the multiverse notion, with its unlimited number of bubble universes, does present a major theoretical problem: it seems to erase the ability of the theory to make predictions—a central requirement of any useful theory. In the words of Alan Guth of the Massachusetts Institute of Technology, one of the creators of inflation theory, "in an eternally inflating universe, anything that can happen will happen; in fact, it will happen an infinite number of times."

In a single universe where events occur a finite number of times, scientists can calculate the relative probability of one event occurring versus another by comparing the number of times these events happen. Yet in a multiverse where everything happens an infinite number of times, such counting is not possible, and nothing is more likely to occur than anything else. One can make any prediction one wants, and it is bound to come true in some universe, but that fact tells you nothing about what will go on in our specific world.

This apparent loss of predictive power has long troubled physicists. Some researchers, including me, have now realized that quantum theory—which, in contrast to the multiverse notion, is concerned with the very smallest particles in existence—may, ironically, point the way to a solution. Specifically, the cosmological picture of the eternally inflating multiverse may be mathematically equivalent to the "many worlds" interpretation of quantum mechanics, which attempts to explain how particles can seem to be in many places at once. As we will see, such a connection between the theories does not just solve the prediction problem; it may also reveal surprising truths about space and time.

Quantum Many Worlds

I came to the idea of a correspondence between the two theories after I revisited the tenets of the many-worlds interpretation of quantum mechanics. This concept arose to make sense of some of the stranger aspects of quantum physics. In the quantum world—a nonintuitive place—cause and effect work differently than they do in the macro world, and the outcome of any process is always probabilistic. Whereas in our macroscopic experience, we can predict where a ball will land when it is thrown based on its starting point, speed and other factors, if that ball were a quantum particle, we could only ever say it has a certain chance of ending up here and another chance of ending up there. This probabilistic nature cannot be avoided by knowing more about the ball, the air currents or such details; it is an intrinsic property of the quantum realm. The same exact ball thrown under the same exact conditions will sometimes land at point A and other times at point B. This conclusion may seem strange, but the laws of quantum mechanics have been confirmed by innumerable experiments and truly describe how nature works at the scale of subatomic particles and forces.

Is There More Than One Universe?

In the quantum world, we say that after the ball is thrown, but before we look for its landing spot, it is in a so-called superposition state of outcomes A and B—that is, it is neither at point A nor at point B but located in a probabilistic haze of *both* points A and B (and many other locations as well). Once we look, however, and find the ball in a certain place—say, point A—then anyone else who examines the ball will also confirm that it sits at A. In other words, before any quantum system is measured, its outcome is uncertain, but afterward all subsequent measurements will find the same result as the first.

In the conventional understanding of quantum mechanics, called the Copenhagen interpretation, scientists explain this shift by saying that the first measurement changed the state of the system from a superposition state to the state A. But although the Copenhagen interpretation does predict the outcomes of laboratory experiments, it leads to serious difficulties at the conceptual level. What does the "measurement" really mean, and why does it change the state of the system from a superposition of possibilities to a single certainty? Does the change of state occur when a dog or even a fly observes the system? What about when a molecule in the air interacts with the system, which we expect to be occurring all the time yet which we do not usually treat as a measurement that can interfere with the outcome? Or is there some special physical significance in a human consciously learning the state of the system?

In 1957 Hugh Everett, then a graduate student at Princeton University, developed the many-worlds interpretation of quantum mechanics that beautifully addresses this issue—although at the time many received it with ridicule, and the idea is still less favored than the Copenhagen interpretation. Everett's key insight was that the state of a quantum system reflects the state of the *whole* universe around it, so that we must include the observer in a complete description of the measurement. In other words, we cannot consider the ball, the wind and the hand that throws it in isolation—we must also include in the fundamental description the person who comes along to inspect its landing spot, as well as everything else in the

Section 2: Time and Space

cosmos at that time. In this picture, the quantum state after the measurement is still a superposition—a superposition of not just two landing spots but two entire worlds! In the first world, the observer finds that the state of the system has changed to A, and therefore any observer in this particular world will obtain result A in all subsequent measurements. But when the measurement was made, another universe split off from the first in which the observer finds, and keeps finding, that the ball landed at point B. This feature explains why the observer—let us say it is a man—thinks that his measurement changes the state of the system; what actually happens is that when he makes a measurement (interacts with the system), he himself divides into two different people who live in two different parallel worlds corresponding to two separate outcomes, A and B.

According to this picture, humans making measurements have no special significance. The state of the entire world continuously branches into many possible parallel worlds that coexist as a superposition. A human observer, being a part of nature, cannot escape from this cycle—the observer keeps splitting into many observers living in many possible parallel worlds, and all are equally "real." An obvious but important implication of this picture is that everything in nature obeys the laws of quantum mechanics, whether small or large.

What does this interpretation of quantum mechanics have to do with the multiverse discussed earlier, which seems to exist in a continuous real space rather than as parallel realities? In 2011 I argued that the eternally inflating multiverse and quantum-mechanical many worlds à la Everett are the same concept in a specific sense. In this understanding, the infinitely large space associated with eternal inflation is a kind of "illusion"—the many bubble universes of inflation do not all exist in a single real space but represent the possible different branches on the probabilistic tree. Around the same time that I made this proposal, Raphael Bousso of the University of California, Berkeley, and Leonard Susskind of Stanford University put forth a similar idea. If true, the many-worlds interpretation of the multiverse would mean that the laws of quantum mechanics do not operate solely in the microscopic realm—they also play a crucial role in determining the global structure of the multiverse even at the largest distance scales.

Black Hole Quandary

To better explain how the many-worlds interpretation of quantum mechanics could describe the inflationary multiverse, I must digress briefly to talk about black holes. Black holes are extreme warps in spacetime whose powerful gravity prevents objects that fall into them from escaping. As such, they provide an ideal testing ground for physics involving strong quantum and gravitational effects. A particular thought experiment about these entities reveals where the traditional way of thinking about the multiverse goes off track, thereby making prediction impossible.

Suppose we drop a book into a black hole and observe from the outside what happens. Whereas the book itself can never escape the black hole, theory predicts that the information in the book will not be lost. After the book has been shredded by the black hole's gravity and after the black hole itself has gradually evaporated by emitting faint radiation (a phenomenon known as Hawking radiation, discovered by physicist Stephen Hawking of the University of Cambridge), outside observers can reconstruct all the information contained in the initial book by closely examining the radiation released. Even before the black hole has completely evaporated, the book's information starts to slowly leak out via each piece of Hawking radiation.

Yet a puzzling thing occurs if we think about the same situation from the viewpoint of someone who is falling into the black hole along with the book. In this case, the book seems to simply pass through the boundary of the black hole and stay inside. Thus, to this inside observer, the information in the book is also contained within the black hole forever. On the other hand, we have just argued that from a distant observer's point of view, the information will be *outside*. Which is correct? You might think that the information is simply duplicated: one copy inside and the other outside. Such a solution, however, is impossible. In quantum mechanics, the so-called no-cloning theorem prohibits faithful, full copying of information. Therefore, it seems that the two pictures seen by the two observers cannot both be true.

Physicists Gerard 't Hooft of Utrecht University in the Netherlands, Susskind and their collaborators have proposed the following solution: the two pictures can both be valid but not at the same time. If you are a distant observer, then the information is outside. You need not describe the interior of the black hole, because you can never access it even in principle; in fact, to avoid cloning information, you must think of the interior spacetime as nonexistent. In contrast, if you are an observer falling into the hole, then the interior is all you have, and it contains the book and its information. This view, however, is possible only at the cost of ignoring the Hawking radiation being emitted from the black hole—but such a conceit is allowed because you yourself have crossed the black hole boundary and accordingly are trapped inside, cut off from the radiation emitted from the boundary. There is no inconsistency in either of these two viewpoints; only if you artificially "patch" the two—which you can never physically do, given that you cannot be both a distant and a falling observer at the same time—does the apparent inconsistency of information cloning occur.

Cosmological Horizons

This black hole conundrum may seem unrelated to the issue of how the many-worlds notion of quantum mechanics and the multiverse can be connected, but it turns out that the boundary of a black hole is similar in important ways to the so-called cosmological horizon—the boundary of the spacetime region within which we can receive signals from deep space. The horizon exists because space is expanding exponentially, and objects farther than this cutoff are receding faster than the speed of light, so any message from them can never reach us. The situation, therefore, is akin to a black hole

Is There More Than One Universe?

viewed by a distant observer. Also, as in the case of the black hole, quantum mechanics requires an observer inside the horizon to view spacetime on the other side of the boundary—in this case, the exterior of the cosmological horizon—as nonexistent. If we consider such spacetime in addition to the information that can be retrieved from the horizon later (analogous to Hawking radiation in the black hole case), then we are overcounting the information. This problem implies that any description of the quantum state of the universe should include only the region within (and on) the horizon—in particular, there can be no infinite space in any single, consistent description of the cosmos.

If a quantum state reflects only the region within the horizon, then where is the multiverse, which we thought existed in an eternally inflating infinite space? The answer is that the creation of bubble universes is probabilistic, like any other process in quantum mechanics. Just as a quantum measurement could spawn many different results distinguished by their probability of occurring, inflation could produce many different universes, each with a different probability of coming into being. In other words, the quantum state representing eternally inflating space is a superposition of worlds—or branches—representing different universes, with each of these branches including only the region within its own horizon.

Because each of these universes is finite, we avoid the problem of predictability that was raised by the prospect of an infinitely large space that encompasses all possible outcomes. The multiple universes in this case do not all exist simultaneously in real space—they coexist only in "probability space," that is, as possible outcomes of observations made by people living inside each world. Thus, each universe—each possible outcome—retains a specific probability of coming into being.

This picture unifies the eternally inflating multiverse of cosmology and Everett's many worlds. Cosmic history then unfolds like this: the multiverse starts from some initial state and evolves into a superposition of many bubble universes. As time passes, the states representing each of these bubbles further branch into more superpositions of states representing the various possible outcomes of "experiments" performed within those universes (these need not be scientific experiments—they can be any physical processes). Eventually the state representing the whole multiverse will thus contain an enormous number of branches, each of which represents a possible world that may arise from the initial state. Quantum-mechanical probabilities therefore determine outcomes in cosmology and in microscopic processes. The multiverse and quantum many worlds are really the same thing; they simply refer to the same phenomenon—superposition—occurring at vastly different scales.

In this new picture, our world is only one of all possible worlds that are allowed by the fundamental principles of quantum physics and that exist simultaneously in probability space.

The Realm Beyond

To know if this idea is correct, we would want to test it experimentally. But is that feasible? It turns out that discovery of one particular phenomenon would lend support to the new thinking. The multiverse could lead to a small amount of negative spatial curvature in our universe—in other words, objects would travel through space not along straight lines as in a flat cosmos but along curves, even in the absence of gravity. Such curvature could happen because, even though the bubble universes are finite as seen from the perspective of the entire multiverse, observers inside a bubble would perceive their universe to be infinitely large, which would make space seem negatively curved (an example of negative curvature is the surface of a saddle, whereas the surface of a sphere is positively curved). If we were inside one such bubble, space should likewise appear to us to be bent.

Evidence so far indicates that the cosmos is flat, but experiments studying how distant light bends as it travels through the cosmos are likely to improve measures of the curvature of our universe by about two orders of magnitude in the next few decades. If these experiments find any amount of negative curvature, they will support the multiverse concept because, although such curvature is technically possible in a single universe, it is implausible there. Specifically, a discovery supports the quantum multiverse picture described here because it can naturally lead to curvature large enough to be detected, whereas the traditional inflationary picture of the multiverse tends to produce negative curvature many orders of magnitude smaller than we can hope to measure.

Interestingly, the discovery of positive curvature would falsify the multiverse notion presented here because inflation theory suggests that bubble universes could produce only negative curvature. On the other hand, if we are lucky, we may even see dramatic signs of a multiverse—such as a remnant from a "collision" of bubble universes in the sky, which may be formed in a single branch in the quantum multiverse. Scientists are, however, far from certain that we will ever detect such signals.

I and other physicists are also pursuing the quantum multiverse idea further on a theoretical level. We can ask fundamental questions such as: How can we determine the quantum state of the entire multiverse? What is time, and how does it emerge? The quantum multiverse picture does not immediately answer these questions, but it does provide a framework to address them. Lately, for instance, I have found that constraints imposed by the mathematical requirement that our theory must include rigorously defined probabilities may enable us to determine the unique quantum state of the entire multiverse. These constraints also suggest that the overall quantum state stays constant even though a physical observer, who is a part of the multiverse state, will see that new bubbles constantly form. This implies that our sense of the universe changing over time and, indeed, the concept of time itself may be an illusion. Time, according to this notion, is an "emergent concept" that arises from a more fundamental reality and seems to exist only within local branches of the multiverse.

Many of the ideas I have discussed are still quite speculative, but it is thrilling that physicists can talk about such big and deep questions based on theoretical progress. Who knows where these explorations will finally lead us? It seems clear, though, that we live in an exciting era in which our scientific explorations reach beyond what we thought to be the entire physical world—our universe—into a potentially limitless realm.

Referenced

Physical Theories, Eternal Inflation, and the Quantum Universe. Yasunori Nomura in *Journal of High Energy Physics*, Vol. 2011, No. 11, Article No. 063; November 2011. Preprint available at https://arxiv.org/abs/1104.2324

Multiverse Interpretation of Quantum Mechanics. Raphael Bousso and Leonard Susskind in *Physical Review D*, Vol. 85, No. 4, Article No. 045007. Published online February 6, 2012. Preprint available at https://arxiv.org/abs/1105.3796

About the Author

Yasunori Nomura is a professor of physics and director of the Berkeley Center for Theoretical Physics at the University of California, Berkeley. He is also a senior faculty scientist at Lawrence Berkeley National Laboratory and a principal investigator at the University of Tokyo's Kavli Institute for the Physics and Mathematics of the Universe.

The Difficult Birth of the "Many Worlds" Interpretation of Quantum Mechanics

By Adam Becker

Over several rounds of sherry late one night in the fall of 1955, the Danish physicist Aage Petersen debated the mysteries at the heart of quantum physics with two graduate students, Charles Misner and Hugh Everett, at Princeton University. Petersen was defending the ideas of his mentor, Niels Bohr, who was the originator of the "Copenhagen interpretation," the standard way of understanding quantum physics. The Copenhagen interpretation, named after the home of Bohr's famous institute, stated that the quantum world of the ultra-tiny was wholly separate from the ordinary world of our everyday experiences.

Quantum physics, Petersen said, applied only to that ultra-tiny realm, where individual subatomic particles performed their strange tricks. It could never be used to describe the world of people and chairs and other objects composed of trillions of trillions of those particles—that world could only be described by the classical physics of Isaac Newton. And, Petersen claimed, this was itself dictated by quantum physics: the mathematics of quantum physics reduced to the mathematics of Newton's physics once the number of particles involved became large.

But Everett incisively attacked the orthodox position advocated by Petersen with alcohol-fueled bravado. Quantum physics, Everett pointed out, didn't really reduce to classical physics for large numbers of particles. According to quantum physics, even normal-sized objects like chairs could be located in two totally separate places at once—a Schrödinger's-cat-like situation known as a "quantum superposition." And, Everett continued, it wasn't right to appeal to classical physics to save the day, because quantum physics was supposed to be a more fundamental theory, one that underpinned classical physics.

Later, in the cold light of day, Everett reconsidered his position—and decided to double down on it. He expanded on his arguments from that evening and turned his attack on the quantum orthodoxy into a PhD thesis. "The time has come ... to treat [quantum physics] in its own right as a fundamental theory without any dependence on classical physics," he wrote in a letter to Petersen.

To solve the problem of superposition, Everett proposed something truly radical, seemingly more appropriate for the pulp sci-fi novels he read in his spare time: he said that quantum physics actually implied an infinite number of near-identical parallel universes, continually splitting off from each other whenever a quantum experiment was performed. This bizarre idea that Everett found lurking in the mathematics of quantum physics came to be known as the "many-worlds" interpretation.

The many-worlds interpretation hit a roadblock almost immediately in the person of Everett's PhD advisor at Princeton, the eminent physicist John Wheeler. Wheeler was a physicist's physicist; he wasn't terribly well known outside of the field, but he knew absolutely everyone important within it. He was a protégé of Bohr, and had also been close with Einstein. Fifteen years before Everett showed up at his door, Wheeler had supervised the PhD of a young Richard Feynman; he would later go on to supervise the PhDs of dozens more renowned physicists (including Kip Thorne, one of the winners of last year's Nobel Prize in Physics).

Everett's strange ideas were initially appealing to Wheeler, because they held the promise of applying quantum theory to the entire universe itself, something that Wheeler desperately wanted to do. But Wheeler was a political animal, and he was wary of incurring the wrath of his mentor Bohr by straying from the quantum orthodoxy preached in Copenhagen. Wheeler's attempt to square this circle was bracingly straightforward: he traveled to Copenhagen to attempt to get Bohr's blessing on Everett's work as an extension of the official Copenhagen line on the nature of the quantum theory.

Is There More Than One Universe?

It didn't go well. Writing to Everett from Copenhagen, Wheeler said that resolving Bohr's criticisms of Everett's ideas "is going to take a lot of time, a lot of heavy arguments with a practical tough-minded man like Bohr, and a lot of writing and rewriting." Wheeler implored Everett to come to Copenhagen himself and "fight with the greatest fighter [i.e. Bohr]."

Everett wasn't particularly interested in fighting or rewriting anything. He was confident in his ideas and was not enticed by the intellectual charms of an academic career. He was more interested in money and the things it could bring: fine food and drink, material luxury, and women. He wanted a *Mad Men* lifestyle, not a professor's office. Everett had already lined up a job that promised to give him just that by the time Wheeler's letter arrived: he had taken a job as a researcher with the Pentagon, gaming out the consequences of hypothetical nuclear missile exchanges at the height of the Cold War.

When Wheeler returned from Europe, he forced Everett to heavily revise his thesis and remove almost all mention of "splitting worlds." Once that was done, Everett left Princeton and never returned to academia. In his later career working for the Pentagon, Everett considered the more grisly possible worlds of nuclear war, and went on to co-author one of the earliest and most influential reports on the subject of radioactive fallout.

But Everett did eventually make it to Copenhagen. In March 1959, he went to Denmark and presented his ideas to Bohr while he was in Europe on other business. As Everett later described it, the meeting was "doomed from the beginning." Neither Everett nor Bohr was swayed. "Bohr's view of quantum mechanics was essentially totally accepted throughout the world by thousands of physicists doing it every day," said Misner, who was also in Copenhagen at the time. "To expect that on the basis of a one-hour talk by a kid he was going to totally change his viewpoint would be unrealistic."

Everett's work fell into deep obscurity. It wasn't revived until the 1970s, and even then, it was slow to catch on. Everett did make one last foray into the academic debate over his work; Wheeler and his colleague Bryce DeWitt invited Everett to speak about his work at the University of Texas in 1977. One of the young physicists in Austin at the time was David Deutsch, who later became a staunch advocate of the many-worlds interpretation. Everett "was full of nervous energy, high-strung, extremely smart," Deutsch recalled. "He was extremely enthusiastic about many universes, and very robust as well as subtle in its defense."

The work of DeWitt, Deutsch, and others led the many-worlds interpretation to become much more popular over the ensuing decades. But Everett didn't live to see the many-worlds interpretation achieve its current status as the most prominent rival to the Copenhagen interpretation. He died of a massive heart attack in 1982, at the age of 51. In accordance with his wishes, his family had him cremated, and left his ashes out to be collected with the trash. But Everett's argumentative and puckish spirit lives on in his theory, born in a drunken debate over 60 years ago, and still inspiring fiery disputes among physicists today.

The views expressed are those of the author(s) and are not necessarily those of Scientific American.

About the Author

Adam Becker is a freelance science journalist and author of What Is Real?, *about the sordid untold history of quantum physics. His writing has appeared in the New York Times, the BBC, NPR and elsewhere. He holds a Ph.D. in cosmology from the University of Michigan.*

Black Hole Discovery Helps to Explain Quantum Nature of the Cosmos

By Edgar Shaghoulian

Where did the universe come from? Where is it headed? Answering these questions requires that we understand physics on two vastly different scales: the cosmological, referring to the realm of galaxy superclusters and the cosmos as a whole, and the quantum—the counterintuitive world of atoms and nuclei.

For much of what we would like to know about the universe, classical cosmology is enough. This field is governed by gravity as dictated by Einstein's general theory of relativity, which doesn't concern itself with atoms and nuclei. But there are special moments in the lifetime of our universe—such as its infancy, when the whole cosmos was the size of an atom—for which this disregard for small-scale physics fails us. To understand these eras, we need a quantum theory of gravity that can describe both the electron circling an atom and Earth moving around the sun. The goal of quantum cosmology is to devise and apply a quantum theory of gravity to the entire universe.

Quantum cosmology is not for the faint of heart. It is the Wild West of theoretical physics, with little more than a handful of observational facts and clues to guide us. Its scope and difficulty have called out to young and ambitious physicists like mythological sirens, only to leave them foundering. But there is a palpable feeling that this time is different and that recent breakthroughs from black hole physics—which also required understanding a regime where quantum mechanics and gravity are equally important—could help us extract some answers in quantum cosmology. The fresh optimism was clear at a recent virtual physics conference I attended, which had a dedicated discussion session about the crossover between the two fields. I expected this event to be sparsely attended, but instead many of the luminaries in physics were there, bursting with ideas and ready to get to work.

Event Horizons

The first indication that there is any relation between black holes and our universe as a whole is that both manifest "event horizons"—points of no return beyond which two people seemingly fall out of contact forever. A black hole attracts so strongly that at some point even light—the fastest thing in the universe—cannot escape its pull. The boundary where light becomes trapped is thus a spherical event horizon around the center of the black hole.

Our universe, too, has an event horizon—a fact confirmed by the stunning and unexpected discovery in 1998 that not only is space expanding, but its expansion is *accelerating*. Whatever is causing this speedup has been called dark energy. The acceleration traps light just as black holes do: as the cosmos expands, regions of space repel one another so strongly that at some point not even light can overcome the separation. This inside-out situation leads to a spherical cosmological event horizon that surrounds us, leaving everything beyond a certain distance in darkness. There is a crucial difference between cosmological and black hole event horizons, however. In a black hole, spacetime is collapsing toward a single point—the singularity. In the universe at large, all of space is uniformly growing, like the surface of a balloon that is being inflated. This means that creatures in faraway galaxies will have their own distinct spherical event horizons, which surround them instead of us. Our current cosmological event horizon is about 16 billion light-years away. As long as this acceleration continues, any light emitted today that is beyond that distance will never reach us. (Cosmologists also speak of a particle horizon, which confusingly is often called a cosmological horizon as well. This refers to the distance beyond which light emitted in the early universe has not yet had time to reach us here on Earth. In our tale, we will be concerned only with the cosmological event horizon, which we will often just call the cosmological horizon. These are unique to universes that accelerate, like ours.)

The similarities between black holes and our universe don't end there. In 1974 Stephen Hawking showed that black holes are not completely black: because of quantum mechanics, they have a temperature and therefore emit matter and radiation, just as all thermal bodies do. This emission, called Hawking radiation, is what causes black holes to eventually evaporate away. It turns out that cosmological horizons also have a temperature and emit matter and radiation because of a very similar effect. But because cosmological horizons surround us and the radiation falls inward, they reabsorb their own emissions and therefore do not evaporate away like black holes.

Hawking's revelation posed a serious problem: if black holes can disappear, so can the information contained within them—which is against the rules of quantum mechanics. This is known as the black hole information paradox, and it is a deep puzzle complicating the quest to combine quantum mechanics and gravity. But in 2019 scientists made dramatic progress. Through a confluence of conceptual and technical advances, physicists argued that the information inside a black hole can actually be accessed from the Hawking radiation that leaves the black hole.

This discovery has reinvigorated those of us studying quantum cosmology. Because of the mathematical similarities between black holes and cosmological horizons, many of us have long believed that we couldn't understand the latter without understanding the former. Figuring out black holes became a warm-up problem—one of the hardest of all time. We haven't fully solved our warm-up problem yet, but now we have a new set of technical tools that provide beautiful insight into the interplay of gravity and quantum mechanics in the presence of black hole event horizons.

Entropy and the Holographic Principle

Part of the recent progress on the black hole information paradox grew out of an idea called the holographic principle, put forward in the 1990s by Gerard 't Hooft of Utrecht University in the Netherlands and Leonard Susskind of Stanford University. The holographic principle states that a theory of quantum gravity that can describe black holes should be formulated not in the ordinary three spatial dimensions that all other physical theories use but instead in two dimensions of space, like a flat piece of paper. The primary argument for this approach is quite simple: a black hole has an entropy—a measure of how much stuff you can stick inside it—that is proportional to the two-dimensional area of its event horizon.

Contrast this with the entropy of a more traditional system—say, a gas in a box. In this case, the entropy is proportional to the three-dimensional volume of the box, not the area. This is natural: you can stick something at every point in space inside the box, so if the volume grows, the entropy grows. But because of the curvature of space within black holes, you can actually increase the volume without affecting the area of the horizon, and this will not affect the entropy! Even though it naively seems you have three dimensions of space to stick stuff in, the black hole entropy formula tells you that you have only two dimensions of space, an area's worth. So the holographic principle says that because of the presence of black holes, quantum gravity should be formulated as a more prosaic nongravitational quantum system in fewer dimensions. At least then the entropies will match.

The idea that space might not be truly three-dimensional is rather compelling, philosophically. At least one dimension of it might be an emergent phenomenon that arises from its deeper nature rather than being explicitly hardwired into the fundamental laws. Physicists who study space now understand that it can emerge from a large collection of simple constituents, similar to other emergent phenomena such as consciousness, which seems to arise from basic neurons and other biological systems.

One of the most exciting aspects of the progress in the black hole information paradox is that it points toward a more general understanding of the holographic principle, which previously had been made precise only in situations very different from our real universe. In the calculations from 2019, however, the way the information inside of the black hole is encoded in the Hawking radiation is mathematically analogous to how a gravitational system is encoded in a lower-dimensional nongravitational system according to the holographic principle. And these techniques can be used in situations more like our universe, giving a potential avenue for understanding the holographic principle in the real world. A remarkable fact about cosmological horizons is that they also have an entropy, given by the exact same formula as the one we use for black holes. The physical interpretation of this entropy is much less clear, and many of us hope that applying the new techniques to our universe will shed light on this mystery. If the entropy is measuring how much stuff you can stick beyond the horizon, as with black holes, then we will have a sharp bound on how much stuff there can be in our universe.

Outside Observers

The recent progress on the black hole information paradox suggests that if we collect all the radiation from a black hole as it evaporates, we can access the information that fell inside the black hole. One of the most important conceptual questions in cosmology is whether the same is possible with cosmological event horizons. We think they radiate like black holes, so can we access what is beyond our cosmological event horizon by collecting its radiation? Or is there some other way to reach across the horizon? If not, then most of our vast, rich universe will eventually be lost forever. This is a grim image of our future—we will be left in the dark.

Almost all attempts to get a handle on this question have required physicists to artificially extricate themselves from the accelerating universe and imagine viewing it from the outside. This is a crucial simplifying assumption, and it more closely mimics a black hole, where we can cleanly separate the observer from the system simply by placing the observer far away. But there seems to be no escaping our cosmological horizon; it surrounds us, and it moves if we move, making this problem much more difficult. Yet if we want to apply our new tools from the study of black holes to the problems of cosmology, we must find a way to look at the cosmic horizon from the outside.

Section 2: Time and Space

There are different ways to construct an outsider view. One of the simplest is to consider a hypothetical auxiliary universe that is quantum-mechanically entangled with our own and investigate whether an observer in the auxiliary universe can access the information in our cosmos, which is beyond the observer's horizon. In work I did with Thomas Hartman and Yikun Jiang, both at Cornell University, we constructed examples of auxiliary universes and other scenarios and showed that the observer can access information beyond the cosmological horizon in the same way that we can access information beyond the black hole horizon. (A complementary paper by Yiming Chen of Princeton University, Victor Gorbenko of EPFL in Switzerland and Juan Maldacena of the Institute for Advanced Study [IAS] in Princeton, N.J., showed similar results.)

But these analyses all have one serious deficiency: when we investigated "our" universe, we used a model universe that is *contracting* instead of expanding. Such universes are much simpler to describe in the context of quantum cosmology. We don't completely understand why, but it's related to the fact that we can think of the interior of a black hole as a contracting universe where everything is getting squished together. In this way, our newfound understanding of black holes could easily help us study this type of universe

Even in these simplified situations, we are puzzling our way through some confusing issues. One problem is that it's easy to construct multiple simultaneous outsider views so that each outsider can access the information in the contracting universe. But this means multiple people can reach the same piece of information and manipulate it independently. Quantum mechanics, however, is exacting: not only does it forbid information from being destroyed, it also forbids information from being replicated. This idea is known as the no-cloning theorem, and the multiple outsiders seem to violate it. In a black hole, this isn't a problem, because although there can still be many outsiders, it turns out that no two of them can independently access the same piece of information in the interior. This limit is related to the fact that there is only one black hole and therefore just one event horizon. But in an expanding spacetime, different observers have different horizons. Recent work that Adam Levine of the Massachusetts Institute of Technology and I did together, however, suggests that the same technical tools from the black hole context work to avoid this inconsistency as well.

Toward a Truer Theory

Although there has been exciting progress, so far we have not been able to directly apply what we learned about black hole horizons to the cosmological horizon in our universe because of the differences between these two types of horizons.

The ultimate goal? No outsider view, no contracting universe, no work-arounds: we want a complete quantum theory of our expanding universe, described from our vantage point within the belly of the beast. Many physicists believe our best bet is to come up with a holographic description, meaning one using fewer than the usual three dimensions of space. There are two ways we can do this. The first is to use tools from string theory, which treats the elementary particles of nature as vibrating strings. When we configure this theory in exactly the right way, we can provide a holographic description of certain black hole horizons. We hope to do the same for the cosmological horizon. Many physicists have put a lot of work into this approach, but it has not yet yielded a complete model for an expanding universe like ours.

The other way to divine a holographic description is by looking for clues from the properties that such a description should have. This approach is part of the standard practice of science—use data to construct a theory that reproduces the data and hope it makes novel predictions as well. In this case, however, the data themselves are also theoretical. They are things we can reliably calculate even without a complete understanding of the full theory, just as we can calculate the trajectory of a baseball without using quantum mechanics. The idea works as follows: we calculate various things in classical cosmology, maybe with a little bit of quantum mechanics sprinkled in, but we try to avoid situations where quantum mechanics and gravity are

Section 2: Time and Space

equally important. This forms our theoretical data. For example, Hawking radiation is a piece of theoretical data. And what must be true is that the full, exact theory of quantum cosmology should be able to reproduce this theoretical datum in an appropriate regime, just as quantum mechanics can reproduce the trajectory of a baseball (albeit in a much more complicated way than classical mechanics).

Leading the charge in extracting these theoretical data is a powerful physicist with a preternatural focus on the problems of quantum cosmology: Dionysios Anninos of King's College London has been working on the subject for more than a decade and has provided many clues toward a holographic description. Others around the world have also joined the effort, including Edward Witten of IAS, a figure who has towered over quantum gravity and string theory for decades but who tends to avoid the Wild West of quantum cosmology. With his collaborators Venkatesa Chandrasekaran of IAS, Roberto Longo of the University of Rome Tor Vergata and Geoffrey Penington of the University of California, Berkeley, he is investigating how the inextricable link between an observer and the cosmological horizon affects the mathematical description of quantum cosmology.

Sometimes we are ambitious and try to calculate theoretical data when quantum mechanics and gravity are equally important. Inevitably we have to impose some rule or guess about the behavior of the full, exact theory in such instances. Many of us believe that one of the most important pieces of theoretical data is the amount and pattern of entanglement between constituents of the theory of quantum cosmology. Susskind and I formulated distinct proposals for how to compute these data, and in hundreds of e-mails exchanged during the pandemic, we argued incessantly over which was more reasonable. Earlier work by Eva Silverstein of Stanford, another brilliant physicist with a long track record in quantum cosmology, and her collaborators provides yet another proposal for computing these theoretical data.

The nature of entanglement in quantum cosmology is a work in progress, but it seems clear that nailing it will be an important step toward a holographic description. Such a concrete, calculable theory is what the subject desperately needs, so that we can compare its outputs with the wealth of theoretical data that are accumulating from scientists. Without this theory, we will be stuck at a stage akin to filling out the periodic table of elements without the aid of quantum mechanics to explain its patterns.

There is a rich history of physicists quickly turning to cosmology after learning something novel about black holes. The story has often been the same: we've been defeated and humbled, but after licking our wounds, we've returned to learn more from what black holes have to teach us. In this instance, the depth of what we've realized about black holes and the breadth of interest in quantum cosmology from scientists around the world may tell a different tale.

About the Author

Edgar Shaghoulian is a theoretical physicist now at the University of California, Santa Cruz. His work focuses on black holes and quantum cosmology.

How the Inside of a Black Hole Is Secretly on the Outside

By Ahmed Almheiri

Theoretical physics has been in crisis mode ever since 1974, when Stephen Hawking argued that black holes destroy information. Hawking showed that a black hole can evaporate, gradually transforming itself and anything it consumes into a featureless cloud of radiation. During the process, information about what fell into the black hole is apparently lost, violating a sacred principle of physics.

This remained an open problem for almost 50 years, but the pieces started falling into place in 2019 through research that I was involved in. The resolution is based on a new understanding of spacetime and how it can be rewired through quantum entanglement, which leads to the idea that part of the inside of a black hole, the so-called island, is secretly on the outside.

To understand how we arrived at these new ideas, we must begin with the inescapable nature of black holes.

A One-Way Street

Nothing seems more hopeless than trying to get out of a black hole—in fact, this impossibility is what defines black holes. They are formed when enough matter is confined within a small enough region that spacetime collapses in on itself in a violent feedback loop of squeezing and stretching that fuels more squeezing and stretching. These tidal forces run to infinity in finite time, marking the abrupt end of an entire region of spacetime at the so-called black hole singularity—the place where time stops and space ceases to make sense.

There is a fine line within the collapsing region that divides the area where escape is possible from the point of no return. This line

Is There More Than One Universe?

is called the event horizon. It is the outermost point from which light barely avoids falling into the singularity. Unless a thing travels faster than light—a physical impossibility—it cannot escape from behind the event horizon; it is irretrievably stuck inside the black hole.

The one-way nature of this boundary is not immediately problematic. In fact, it is a robust prediction of the general theory of relativity. The danger starts when this theory interacts with the wild world of quantum mechanics.

Something Out of Nothing

Quantum theory redeems black holes from being the greedy monsters they are made out to be. Every calorie of energy they consume they eventually give back in the form of Hawking radiation—energy squeezed out of the vacuum near the event horizon.

The idea of getting something out of nothing may sound absurd, but absurdity is not the worst allegation made against quantum mechanics. The emptiness of the vacuum in quantum theory belies a sea of particles—photons, electrons, gravitons, and more—that conspire to make empty space feel empty. These particles come in carefully arranged pairs, acting hand in hand as the glue that holds spacetime together.

Particle pairs that straddle the event horizon of a black hole, however, become forever separated from each other. The newly divorced particles peel away from the horizon in opposite directions, with one member crashing into the singularity and the other escaping the black hole's gravitational pull in the form of Hawking radiation. This process is draining for the black hole, causing it to get lighter and smaller as it emits energy in the form of the outgoing particles. Because of the law that energy must be conserved, the particles trapped inside must then carry negative energy to account for the decrease in the total energy of the black hole.

From the outside, the black hole appears to be burning away (although it happens so slowly, you can't see it happening in real life). When you burn a book, the words on its pages imprint themselves on the pattern of the emanating light and the remaining ashes. This information is thus preserved, at least in principle. If an evaporating black hole were a normal system like the burning book, then the information about what falls into it would be encoded into the emerging Hawking radiation. Unfortunately, this is complicated by the quantum-mechanical relation among the particles across the horizon.

Einstein's Enemy

The issue begins with the end of the pairing of the two particles straddling the event horizon. Despite being separated, they maintain a quantum union that transcends space and time—they are connected by entanglement. Rejected as an absurdity by the physicists who predicted it, quantum entanglement is perhaps one of the weirdest aspects of our universe and arguably one of its most essential. The concept was first concocted by Albert Einstein, Boris Podolsky and Nathan Rosen as a rebuttal against what was then the nascent theory of quantum mechanics. They cited entanglement as a reason the theory must be incomplete—"spooky" is how Einstein famously described the phenomenon.

For a simple example of entanglement, consider two coins in a superposition—the quantum phenomenon of being in multiple states until a measurement is made—of both coins being either heads or tails. The coins aren't facing heads and tails at the same time—that's physically impossible—but the superposition signifies that the chance of observing the pair of coins in either orientation, both heads or both tails, is a probability of one half. There is no chance of ever finding the coins in opposite orientations. The two coins are entangled; the measurement result of one predicts the result of the other with complete certainty. Either coin by itself is completely random, devoid of information, but the randomness of the pair is perfectly correlated.

The scientists were troubled by how the two coins appeared to influence each other without having to come into physical contact.

The coins could be in separate galaxies while still maintaining the same amount of entanglement between them. Einstein was unnerved by the apparent "spooky action at a distance" linking the results of the two separate random measurements.

The irony is that Einstein himself is in a superposition of being both wrong and right. He was right to recognize the importance of entanglement in distinguishing quantum mechanics from classical physics. What he got wrong can be summed up with the truism "correlation does not imply causation." Although the fates of the particles are inextricably correlated, the measurement outcome of one does not cause the outcome of the other. It turns out that quantum mechanics simply allows for a new, higher degree of correlation than we are used to.

Information Lost

Because Hawking radiation is composed of one half of a collection of entangled pairs, it emerges from the black hole in a completely random state—if the particles were coins, they would be observed to be heads or tails with equal probability. Hence, we cannot infer anything useful about the contents of the black hole from the random measurements of the radiation. This means that an evaporating black hole is basically a glorified information shredder, except unlike the mechanical kind, it does a thorough job.

We can measure the lack of information—or the randomness—in the Hawking radiation by thinking about the amount of entanglement between the radiation and the black hole. This is because one member of an entangled pair is always random, and the outside members are all that remains by the end of the evaporation. The calculation of randomness goes by many names, including entanglement entropy, and it grows with every emerging Hawking particle, plateauing at a large value once the black hole has completely disappeared.

This pattern differs from what happens when information is preserved, as in the example of a burning book. In such a case, the entropy may rise initially, but it has to peak and then fall to zero by the end of the process. The intuition behind this rule is clear when you think about a standard deck of cards: suppose you are dealt cards from a 52-card deck, one by one, facing down. The entropy of the cards in your possession is simply a measure of your ignorance of what's on the other side of the cards—specifically, the number of possibilities of what they could be. If you have been dealt just one card, the entropy is 52 because there are 52 possibilities. But as you are dealt more, the entropy rises, peaking at 500 trillion for 26 cards, which could be any of 500 trillion different combinations. After this, though, the possible mixes of cards, and thus the entropy, go back down, reaching 52 again when you have 51 cards. Once you have all the cards, you are certain of exactly what you have—the entire deck—and the entropy is zero. This rising and falling pattern of entropy, known as the Page curve, applies to all normal quantum-mechanical systems. The time at which the entropy peaks and then starts to decrease is the Page time.

The destruction of information inside black holes spells disaster for physics because the laws of quantum mechanics stipulate that information cannot be obliterated. This is the famous information paradox—the fact that a sprinkling of quantum mechanics onto the description of black holes leads to a seemingly insurmountable inconsistency. Physicists knew we needed a more complete understanding of quantum-gravitational physics to generate the Page curve for the Hawking radiation. Unsurprisingly, this task proved difficult.

An Eventful Horizon

Part of the challenge was that no minor tweaking of the evaporation process was sufficient to generate the Page curve and send the entropy back down to zero. What we needed was a drastic reimagining of the structure of a black hole.

In a paper I published in 2013 with Donald Marolf, the late Joseph Polchinski, and Jamie Sully (known collectively as AMPS), we tried out several ways to modify the picture of evaporating black holes using a series of *gedankenexperiments*—the German term for the kind of thought experiments Einstein popularized. Through our trials we concluded that to save the sanctity of information, one of two things had to give: either physics must be nonlocal—allowing for information to instantaneously disappear from the interior and appear outside the event horizon—or a new process must kick in at the Page time. To preclude the increase of entropy, this process would have to break the entanglement between the particle pairs across the event horizon. The former option—making physics nonlocal—was too radical, so we decided to go with the latter.

This modification helps to preserve information, but it poses another paradox. Recall that the entanglement across the horizon was a result of having empty space there—the way the vacuum is maintained by a sea of entangled pairs of particles. The entanglement is key; breaking it comes at the cost of creating a wall of extremely high-energy particles, which our group named the firewall. Having such a firewall at the horizon would forbid anything from entering the black hole. Instead infalling matter would be vaporized on contact. The black hole at the Page time would suddenly lose its interior, and spacetime would come to an end, not at the singularity deep inside the black hole but right there at the event horizon. This conclusion is known as the firewall paradox, a catch-22 that meant any solution to the information paradox must come at the cost of destroying what we know about black holes. If ever there were a quagmire, this would be it.

Fluctuating Wormholes

Eventually my colleagues and I realized that both the information paradox and the newer firewall paradox arose because our attempts to meld quantum mechanics and black hole physics were too timid. It wasn't enough to apply quantum mechanics to only the matter present in black holes—we had to devise a quantum treatment of the black hole spacetime as well. Although quantum effects on spacetime are usually very small, they could be enhanced by the large entanglement produced by the evaporation. Such an effect may be subtle, but its implications would be huge.

To consider the quantum nature of spacetime, we relied on a technique designed by Richard Feynman called the path integral of quantum mechanics. The idea is based on the weird truth that, according to quantum theory, particles don't simply travel along a single path from point A to point B—they travel along *all* the different paths connecting the two points. The path integral is a way of describing a particle's travels in terms of a quantum superposition of all possible routes. Similarly, a quantum spacetime can be in a superposition of different complicated shapes evolving in different ways. For instance, if we start and end with two regular black holes, the quantum spacetime within them has a nonzero probability of creating a short-lived wormhole that temporarily bridges their interiors.

Usually the probability of this happening is vanishingly slim. When we carry out the path integral in the presence of the Hawking radiation of multiple black holes, however, the large entanglement between the Hawking radiation and the black hole interiors amplifies the likelihood of such wormholes. This realization came to me through work I did in 2019 with Thomas Hartman, Juan Maldacena, Edgar Shaghoulian and Amirhossein Tajdini, and it was also the result of an independent collaboration by Geoffrey Penington, Stephen Shenker, Douglas Stanford and Zhenbin Yang.

Islands Beyond the Horizon

Why does it matter if some black holes are connected by wormholes? It turns out that they modify the answer of how much entanglement entropy there is between the black hole and its Hawking radiation. The key is to measure this entanglement entropy in the presence of multiple copies of the system. This is known as the replica trick.

The relevant physical effect of these temporary wormholes is to swap out the interiors among the different black holes. This happens literally: what was in one black hole gets shoved into one of the other copies far away, and the original black hole assumes a new spacetime interior from a different one. The swapped region of the black hole interior is called the island, and it encompasses almost the entire interior up to the event horizon.

The swapping is exactly what the doctor ordered! Focusing on one of the black holes and its Hawking radiation, the swapped-out island takes with it all the partner particles that are entangled with the outgoing Hawking radiation, and hence, technically, there is no entanglement between the black hole and its radiation.

Including this potential effect of wormholes produces a new formula for the entanglement entropy of the radiation when applied to a single copy of the system. Instead of Hawking's original calculation, which simply counts the number of Hawking particles outside a black hole, the new formula curiously treats the island as if it were outside and a part of the exterior Hawking radiation. Therefore, the entanglement between the island and the exterior should not be counted toward the entropy. Instead the entropy that it predicts comes almost entirely from the probability of the swap actually occurring, which is equal to the area of the boundary of the island—roughly the area of the event horizon—divided by Newton's gravitational constant. As the black hole shrinks, this contribution to the entropy decreases. This is the island formula for the entanglement entropy of the Hawking radiation.

The final step in computing the entropy is to take the minimum between the island formula and Hawking's original calculation. This gives us the Page curve that we've been after. Initially we calculate the entanglement entropy of the radiation with Hawking's original formula because the answer starts off smaller than the area of the event horizon of the black hole. But as the black hole evaporates, the area shrinks, and the new formula takes the baton as the true representative of the radiation's entanglement entropy.

What is remarkable about this result is that it solves two paradoxes with one formula. It appears to address the firewall

Section 2: Time and Space

paradox by supporting the option of nonlocality that my AMPS group originally dismissed. Instead of breaking the entanglement at the horizon, we are instructed to treat the inside—the island—as part of the outside. The island itself becomes nonlocally mapped to the outside. And the formula solves the information paradox by revealing how black holes produce the Page curve and preserve information.

Let's take a step back and think about how we got here. The origins of the information paradox can be traced back to the incompatibility between the sequestering of information by the event horizon and the quantum-mechanical requirement of information flow outside the black hole. Naive resolutions of this tension lead to drastic modifications of the structure of black holes; however, subtle yet dramatic effects from fluctuating wormholes change everything. What emerges is a self-consistent picture that lets a black hole retain its regular structure as predicted by general relativity, albeit with the presence of an implicit though powerful nonlocality. This nonlocality is screaming that we should consider a portion of the black hole's interior—the island—as part of the exterior, as a single unit with the outside radiation. Thus, information can escape a black hole not by surmounting the insurmountable event horizon but by simply falling deeper into the island.

Despite the excitement of this breakthrough, we have only begun to explore the implications of spacetime wormholes and the island formula. Curiously, while they ensure that the island is mapped onto the radiation, they do not generate a definite prediction for specific measurements of the Hawking radiation. What they do teach us, however, is that wormholes are the missing ingredient in Hawking's original estimation of the randomness in the radiation and that gravity is in fact smart enough to comply with quantum mechanics.

Through these wormholes, gravity harnesses the power of entanglement to achieve nonlocality, which is just as unnerving to us as the entanglement that originally spooked Einstein. We must admit that, at some level, Einstein was right after all.

About the Author

Ahmed Almheiri is a theoretical physicist at New York University in Abu Dhabi, United Arab Emirates, where he studies the connections between quantum information and quantum gravity.

Section 3: Pulling Strings

3.1 Will String Theory Finally Be Put to the Experimental Test?
By Brendan Z. Foster

3.2 String Theory May Create Far Fewer Universes Than Thought
By Clara Moskowitz and Lee Billings

Will String Theory Finally Be Put to the Experimental Test?

By Brendan Z. Foster

Many physicists consider string theory our best hope for combining quantum physics and gravity into a unified theory of everything. Yet a contrary opinion is that the concept is practically pseudoscience, because it seems to be nearly impossible to test through experiments. Now some scientists say we may have a way to do exactly that, thanks to a new conjecture that pits string theory against cosmic expansion.

What it comes down to is this question: Does the universe show us all of its quantum secrets, or does it somehow hide those details from our classical eyes? Because if the details can be seen, string theory might not be able to explain them.

One way to rule out the idea is if we can prove that it does not predict an essential feature of the universe. And string theory, it turns out, has a persistent problem describing the most popular account of what went on during the universe's earliest moments after the big bang: inflation.

"Inflation is the most compelling explanation for why our universe looks the way it does and where the structure came from," says Marilena Loverde, a physicist at Stony Brook University. Inflation explains how, in a sense, we got everything in the universe from nothing. The theory says that the early universe went through a phase of extreme expansion. The process magnified random blips in the quantum vacuum and converted them into the galaxies and other stuff around us.

Theorists have had difficulty, though, showing how, or if, inflation works in string theory. The most promising road to doing so—the so-called KKLT construction—does not convince everyone. "It depends who you ask," says Suddhasattwa Brahma, a cosmologist at McGill University. "It has been a lingering doubt in the back of the minds of many in string theory: Does it really work?"

In 2018 a group of string theorists took a series of suggestive results and argued that this difficulty reflected an impossibility—that perhaps inflation just cannot happen in the theory. This so-called de Sitter swampland conjecture claimed that any version of the concept that could describe de Sitter space—a term for the kind of universe in which we expect inflation to take place—would have some kind of technical flaw that put it in a "swampland" of rejected theories.

No one has proved the swampland conjecture, and several string theorists still expect that the final form of the theory will have no problem with inflation. But many believe that although the conjecture might not hold up rigidly, something close to it will. Brahma hopes to refine the swampland conjecture to something that would not bar inflation entirely. "Maybe there can be inflation," he says. "But it has to be a very short period of inflation."

Any limit on inflation would raise the prospect of testing string theory against actual data, but a definite test requires a proof of the conjecture. According to Cumrun Vafa, a physicist at Harvard University and one of the swampland conjecture's authors, researchers can start to build a case for the idea if they can connect it to trusted physical laws. "There are two levels of it," he says. "First is being more confident in the principle. And then there's explaining it."

One approach to building confidence might try to explain what sort of physical rule would limit inflation—or, to put the inquiry in a more practical way: How could string theorists hope to persuade cosmologists to reconsider a favored theory?

These kinds of questions led Vafa and his Harvard collaborator Alek Bedroya to seek out a physics-based reason that could justify the swampland conjecture. They found a candidate in a surprising place. It turns out that inflation already has an unsolved problem looking for a solution: theorists have not all agreed on what happens to the very tiniest quantum details when expansion occurs and magnifies the static of the vacuum.

Physicists lack a working theory that describes the world below the level of the so-called Planck length, an extremely minute distance where they expect the quantum side of gravity to appear. Proponents of inflation have typically had to assume that they can one day work those "trans-Planckian" details into it and that they will not make a big difference to any predictions. But how that step will happen remains an open question.

Vafa and Bedroya have given a simple answer: forget about it. Their new trans-Planckian censorship conjecture asserts that extremely tiny quantum fuzziness should always stay extremely tiny and quantum, despite the magnifying effect of expansion. If this idea is true, it implies limits on the amount of inflation that could happen, because too much of it would mean too much magnification of the trans-Planckian details.

So in a new twist for string theory, researchers can actually look to the sky for some answers. How much inflation is too much for the censorship conjecture? The situation is a bit complicated. Several different models for the actual process of inflation exist, and astrophysicists do not yet have data that confirm any one of them, or the basic idea as a whole, as the correct description of our universe. Researchers have begun working out the limits the new conjecture puts on the many versions of inflation. Some have a built-in way to hide trans-Planckian details, but Loverde says that many of the typical models conflict with the conjecture.

One clear conflict comes from "primordial" gravitational waves. These waves, which theorists expect arise during the inflationary phase, would have left behind a faint but distinct sign in the cosmic microwave background. So far, they have not been seen, but telescopes are actively looking for them. The censorship conjecture would only allow a "ridiculously, unobservably small" amount, Loverde says—so small that any sign of these waves would mean the conjecture does not apply to our universe unless theorists can come up with a different explanation for them.

Does this conjecture really amount to a test of string theory? No, it is too early to say that, according to Vafa. The principles are still just conjectures—for now. "The more one connects these principles together—surprising, unexpected relations—the more it becomes believable why it's true," he says.

About the Author

Brendan Z. Foster is a science writer and audio journalist based in Saint Paul, Minn. He has a Ph.D. in physics from the University of Maryland.

String Theory May Create Far Fewer Universes Than Thought

By Clara Moskowitz and Lee Billings

The problem with string theory, according to some physicists, is that it makes too many universes. It predicts not one but some 10,500 versions of spacetime, each with its own laws of physics. But with so many universes on the table, how can the theory explain why ours has the features it does?

Now some theorists suggest most—if not all—of those universes are actually forbidden, at least if we want them to have stable dark energy, the supposed force accelerating the expansion of the cosmos. To some, eliminating so many possible universes is not a drawback but a major step forward for string theory, offering new hope of making testable predictions. But others say the multiverse is here to stay, and the proposed problem with all those universes is not a problem at all.

The debate was a hot topic in the summer of 2018 in Japan, where string theorists convened for the conference Strings 2018. "This is really something new and it's led to a controversy within the field," says Ulf Danielsson, a physicist at Uppsala University in Sweden. The conversation centers on a pair of papers posted on the preprint server arXiv in June 2018 taking aim at the so-called landscape of string theory—the incomprehensible number of potential universes that result from the many different solutions to string theory's equations that produce the ingredients of our own cosmos, including dark energy. But the vast majority of the solutions found so far are mathematically inconsistent, the papers contend, putting them not in the landscape but in the so-called swampland of universes that cannot actually exist. Scientists have known many solutions must fall in this swampland for years, but the idea that most, or maybe all, of the landscape solutions might live there would be a major change. In fact, it may be theoretically impossible to find a valid solution to string theory that includes stable dark energy, says Cumrun Vafa, a Harvard University physicist who led the work on the two papers.

Lost in the Multiverse

String theory is an attempt to describe the whole universe under a single "theory of everything" by adding extra dimensions of spacetime and thinking of particles as minuscule vibrating loops. Many string theorists contend it is still the most promising direction for pursuing Albert Einstein's dream of uniting his general theory of relativity with the conflicting microscopic world of quantum mechanics. Yet the notion of a string theory landscape that predicts not just one universe but many has put some physicists off. "If it's really the landscape, in my view it's death for the theory because it loses all predictive value," says Princeton University physicist Paul Steinhardt, who collaborated on one of the recent papers. "Literally anything is possible." To Steinhardt and others, the newfound problems with dark energy offer string theory a way out. "This picture with a big multiverse could be mathematically wrong," Danielsson says. "Paradoxically this makes things much more interesting because that means string theory is much more predictive than we thought it was."

Some string theorists such as Savdeep Sethi of the University of Chicago welcome the reevaluation that is happening now. "I think this is exciting," he says. "I've been a skeptic of the landscape for a long time. I'm really happy to see the paradigm shift away from this belief that we have this proven set of solutions." But not everyone buys the argument that the landscape actually belongs in the swampland—especially the research team that established one of the earliest versions of the landscape in the first place back in 2003, which goes by the acronym KKLT after the scientists' last names. "I think it's very healthy to make these conjectures and check what other things could be going on, but I don't see either theoretical or experimental reasons to take such a conjecture very seriously," says KKLT member Shamit Kachru of Stanford University. And Eva Silverstein, a Stanford physicist who also helped build the early landscape models, likewise doubts Vafa and his

Is There More Than One Universe?

colleagues' argument. "I think the ingredients KKLT use and the way they put them together is perfectly valid," she says. Juan Maldacena, a theorist at the Institute for Advanced Study, says he also still supports the idea of string theory universes with stable dark energy.

And many theorists are perfectly happy with the string theory multiverse. "It is true that if this landscape picture is correct, the bit of the universe we're in compared to the multiverse will be like our solar system within the universe," Kachru says. And that is a good thing, he adds. Johannes Kepler originally sought a fundamental reason for why Earth lies the distance it does from the sun. But now we know the sun is just one of billions of stars in the galaxy, each with its own planets, and the Earth-sun distance is simply a random number rather than a result of some deep mathematical principle. Likewise, if the universe is one of trillions within the multiverse, the particular parameters of our cosmos are similarly random. The fact that these numbers seem perfectly fine-tuned to create a habitable universe is a selection effect—humans will of course find themselves in one of the rare corners of the multiverse where it is possible for them to have evolved.

The Accelerating Universe

If it is true that string theory cannot accommodate stable dark energy, that may be a reason to doubt string theory. But to Vafa, it is a reason to doubt dark energy—that is, dark energy in its most popular form, called a cosmological constant. The idea originated in 1917 with Einstein and was revived in 1998 when astronomers discovered that not only is spacetime expanding—the rate of that expansion is picking up. The cosmological constant would be a form of energy in the vacuum of space that never changes and counteracts the inward pull of gravity. But it is not the only possible explanation for the accelerating universe. An alternative is "quintessence," a field pervading spacetime that can evolve. "Regardless of whether one can realize a stable dark energy in string theory or not, it turns out that the idea of having dark energy changing over time is actually more natural in string theory," Vafa says.

"If this is the case, then one can measure this sliding of dark energy by astrophysical observations currently taking place."

So far all astrophysical evidence supports the cosmological constant idea, but there is some wiggle room in the measurements. Upcoming experiments such as Europe's Euclid space telescope, NASA's Wide-Field Infrared Survey Telescope (WFIRST) and Chile's Simons Observatory being built in the desert will look for signs that dark energy was stronger or weaker in the past than the present. "The interesting thing is that we're already at a sensitivity level to begin to put pressure on [the cosmological constant theory]," Steinhardt says. "We don't have to wait for new technology to be in the game. We're in the game now." And even skeptics of Vafa's proposal support the idea of considering alternatives to the cosmological constant. "I actually agree that [a changing dark energy field] is a simplifying method for constructing accelerated expansion," Silverstein says. "But I don't think there's any justification for making observational predictions about the dark energy at this point."

Quintessence is not the only other option. In the wake of Vafa's papers, Danielsson and his colleagues proposed another way of fitting dark energy into string theory. In their vision, our universe is the three-dimensional surface of a bubble expanding within a larger-dimensional space. "The physics within this surface can mimic the physics of a cosmological constant," Danielsson says. "This is a different way of realizing dark energy compared to what we've been thinking so far."

A Beautiful Theory

Ultimately the debate going on in string theory centers on a deep question: What is the point of physics? Should a good theory be able to explain the particular characteristics of the universe around us or is that asking too much? And when a theory conflicts with the way we think our universe works, do we abandon the theory or the things we think we know?

String theory is incredibly appealing to many scientists because it is "beautiful"—its equations are satisfying and its proposed explanations elegant. But so far it lacks any experimental evidence supporting it—and even worse, any reasonable prospects for gathering such evidence. Yet even the suggestion that string theory may not be able to accommodate the kind of dark energy we see in the cosmos around us does not dissuade some. "String theory is so rich and beautiful and so correct in almost all the things that it's taught us that it's hard to believe that the mistake is in string theory and not in us," Sethi says. But perhaps chasing after beauty is not a good way to find the right theory of the universe. "Mathematics is full of amazing and beautiful things, and most of them do not describe the world," physicist Sabine Hossenfelder of the Frankfurt Institute for Advanced Studies wrote in her recent book, *Lost in Math: How Beauty Leads Physics Astray* (Basic Books, 2018).

Despite the divergence of opinions, physicists are a friendly bunch and are united by their common goal of understanding the universe. Kachru, one of the founders of the landscape idea, worked with Vafa, the landscape's critic, as his undergraduate adviser—and the two are still friends. "He asked me once if I'd bet my life these [landscape solutions] exist," Kachru says. "My answer was, 'I wouldn't bet my life, but I'd bet his!'"

Referenced

On the Cosmological Implications of the String Swampland. Prateek Agrawal, Georges Obied, Paul J. Steinhard and Cumrun Vafa in *Physics Letters B*, Vol. 784, pages 271–276; September 10, 2018.

Emergent de Sitter Cosmology from Decaying Anti–de Sitter Space. Souvik Banerjee, Ulf Danielsson, Giuseppe Dibitetto, Suvendu Giri and Marjorie Schillo in *Physical Review Letters*, Vol. 121, Article 261301; December 27, 2018.

De Sitter Space and the Swampland. Georges Obied, Hirosi Ooguri, Lev Spodyneiko and Cumrun Vafa (in press). Preprint available at arXiv:1806.08362v3

About the Authors

Clara Moskowitz is Scientific American's *senior editor covering space and physics. She has a bachelor's degree in astronomy and physics from Wesleyan University and a graduate degree in science journalism from the University of California, Santa Cruz.*

Lee Billings is a senior editor for space and physics at Scientific American.

Section 4: Searching for Answers

4.1 Looking for Life in the Multiverse
   By Alejandro Jenkins and Gilad Perez

4.2 Does a Multiverse Fermi Paradox Disprove the Multiverse?
   By Caleb A. Scharf

4.3 Was Our Universe Created in a Laboratory?
   By Avi Loeb

4.4 Multiverse Theories Are Bad for Science
   By John Horgan

4.5 Life Quest: Could Parallel Universes Be Congenial to Life?
   By Mariette DiChristina

4.6 Multiverse Controversy Heats Up Over Gravitational Waves
   By Clara Moskowitz

Looking for Life in the Multiverse

By Alejandro Jenkins, Gilad Perez

The typical Hollywood action hero skirts death for a living. Time and again, scores of bad guys shoot at him from multiple directions but miss by a hair. Cars explode just a fraction of a second too late for the fireball to catch him before he finds cover. And friends come to the rescue just before a villain's knife slits his throat. If any one of those things happened just a little differently, the hero would be *hasta la vista, baby*. Yet even if we have not seen the movie before, something tells us that he will make it to the end in one piece.

In some respects, the story of our universe resembles a Hollywood action movie. Several physicists have argued that a slight change to one of the laws of physics would cause some disaster that would disrupt the normal evolution of the universe and make our existence impossible. For example, if the strong nuclear force that binds together atomic nuclei had been slightly stronger or weaker, stars would have forged very little of the carbon and other elements that seem necessary to form planets, let alone life. If the proton were just 0.2 percent heavier than it is, all primordial hydrogen would have decayed almost immediately into neutrons, and no atoms would have formed. The list goes on.

The laws of physics—and in particular the constants of nature that enter into those laws, such as the strengths of the fundamental forces—might therefore seem finely tuned to make our existence possible. Short of invoking a supernatural explanation, which would be by definition outside the scope of science, a number of physicists and cosmologists began in the 1970s to try solving the puzzle by hypothesizing that our universe is just one of many existing universes, each with its own laws. According to this "anthropic" reasoning, we might just occupy the rare universe where the right conditions happen to have come together to make life possible.

Amazingly, the prevailing theory in modern cosmology, which emerged in the 1980s, suggests that such "parallel universes" may really exist—in fact, that a multitude of universes would incessantly pop out of a primordial vacuum the way ours did in the big bang. Our universe would be but one of many pocket universes within a wider expanse called the multiverse. In the overwhelming majority of those universes, the laws of physics might not allow the formation of matter as we know it or of galaxies, stars, planets and life. But given the sheer number of possibilities, nature would have had a good chance to get the "right" set of laws at least once.

Our recent studies, however, suggest that some of these other universes—assuming they exist—may not be so inhospitable after all. Remarkably, we have found examples of alternative values of the fundamental constants, and thus of alternative sets of physical laws, that might still lead to very interesting worlds and perhaps to life. The basic idea is to change one aspect of the laws of nature and then make compensatory changes to other aspects.

Our work did not address the most serious fine-tuning problem in theoretical physics: the smallness of the "cosmological constant," thanks to which our universe neither recollapsed into nothingness a fraction of a second after the big bang, nor was ripped part by an exponentially accelerating expansion. Nevertheless, the examples of alternative, potentially habitable universes raise interesting questions and motivate further research into how unique our own universe might be.

The Weakless Way of Life

The conventional way scientists find out if one particular constant of nature is finely tuned or not is to turn that "constant" into an adjustable parameter and tweak it while leaving all other constants unaltered. Based on their newly modified laws of physics, the scientists then "play the movie" of the universe—they do calculations, what-if scenarios or computer simulations—to see what disaster occurs first. But there is no reason why one should tweak only one parameter at a time. That situation resembles trying to drive a car by varying only your latitude or only your longitude, but not both: unless you are traveling on a grid, you are destined to run off the road. Instead one can tweak multiple parameters at once.

To search for alternative sets of laws that still give rise to complex structures capable of sustaining life, one of us (Perez) and his collaborators did not make just small tweaks to the known laws of physics: they completely eliminated one of the four known fundamental forces of nature.

By their very name, the fundamental forces sound like indispensable features of any self-respecting universe. Without the strong nuclear force to bind quarks into protons and neutrons and those into atomic nuclei, matter as we know it would not exist. Without the electromagnetic force, there would be no light; there would also be no atoms and no chemical bonds. Without gravity, there would be no force to coalesce matter into galaxies, stars and planets.

The fourth force, the weak nuclear force, has a subtler presence in our everyday life but still has played a major role in the history of our universe. Among other things, the weak force enables the reactions that turn neutrons into protons, and vice versa. In the first instants of the big bang, after quarks (among the first forms of matter to appear) had united in groups of three to form protons and neutrons, collectively called baryons, groups of four protons were then able to fuse together and become helium 4 nuclei, made of two protons and two neutrons. This so-called big bang nucleosynthesis took place a few seconds into the life of our universe, when it was already cold enough for baryons to form but still hot enough for the baryons to undergo nuclear fusion. Big bang nucleosynthesis produced the hydrogen and helium that would later form stars, where nuclear fusion and other processes would forge virtually all other naturally occurring elements. And to this day, the fusion of four protons to make helium 4 continues inside our sun, where it produces most of the energy that we receive from it.

Section 4: Searching for Answers

Without the weak nuclear force, then, it seems unlikely that a universe could contain anything resembling complex chemistry, let alone life. Yet in 2006 Perez's team discovered a set of physical laws that relied on only the other three forces of nature and still led to a congenial universe.

Eliminating the weak nuclear force required several modifications to the so-called Standard Model of particle physics, the theory that describes all forces except gravity. The team showed that the tweaks could be done in such a way that the behavior of the other three forces—and other crucial parameters such as the masses of the quarks—would be the same as in our world. We should stress that this choice was a conservative one, intended to facilitate the calculation of how the universe would unfold. It is quite possible that a wide range of other "weakless" universes exist that are habitable but look nothing like our own.

In the weakless universe, the usual fusing of protons to form helium would be impossible, because it requires that two of the protons convert into neutrons. But other pathways could exist for the creation of the elements. For example, our universe contains overwhelmingly more matter than antimatter, but a small adjustment to the parameter that controls this asymmetry is enough to ensure that the big bang nucleosynthesis would leave behind a substantial amount of deuterium nuclei. Deuterium, also known as hydrogen 2, is the isotope of hydrogen whose nucleus contains a neutron in addition to the usual proton. Stars could then shine by fusing a proton and a deuterium nucleus to make a helium 3 (two protons and one neutron) nucleus.

Such weakless stars would be colder and smaller than the stars in our own universe. According to computer simulations by astrophysicist Adam Burrows of Princeton University, they could burn for about seven billion years—about the current age of our sun—and radiate energy at a rate that would be a few percent of that of the sun.

Next Generation

Just like stars in our universe, weakless stars could synthesize elements as heavy as iron through further nuclear fusion. But the typical reactions that in our stars lead to elements beyond iron would be compromised, primarily because few neutrons would be available for nuclei to capture to become heavier isotopes, the first step in the formation of heavier elements. Small amounts of heavy elements, up to strontium, might still be synthesized inside weakless stars by other mechanisms.

In our universe, supernova explosions disperse the newly synthesized elements into space, and synthesize more of the elements themselves. Supernovae can be of several types: in the weakless universe, the supernova explosions caused by collapsing ultramassive stars would fail, because it is the emission of neutrinos, produced via the weak-force interactions, that transmits energy out of a star's core so as to sustain the shock wave that is causing the explosion. But a different type of supernova—the thermonuclear explosion of a star triggered by accretion, rather than by gravitational collapse—would still take place. Thus, elements could be dispersed into interstellar space, where they could seed new stars and planets.

Given the relative coldness of the weakless stars, a weakless Earth-like body would have to be about six times closer to its sun to stay as warm as our own Earth. To the inhabitants of such a planet, the sun would look much bigger. Weakless Earths would be significantly different from our own Earth in other ways. In our world, plate tectonics and volcanic activity are powered by the radioactive decay of uranium and thorium deep within Earth. Without these heavy elements, a typical weakless Earth might have a comparatively boring and featureless geology—except if gravitational processes provided an alternative source of heating, as happens on some moons of Saturn and Jupiter.

Chemistry, on the other hand, would be very similar to that of our world. One difference would be that the periodic table would stop at iron, except for extremely small traces of other elements. But this

Section 4: Searching for Answers

limitation should not prevent life-forms similar to the ones we know from evolving. Thus, even a universe with just three fundamental forces could be congenial to life.

Another approach, pursued by the other of us (Jenkins) and his collaborators, searches for alternative sets of laws by making smaller tweaks to the Standard Model than in the case of the weakless universe, though still involving multiple parameters at once. In 2008 the group studied to what extent the masses of the three lightest of the six quarks—called the up, down and strange quarks—may vary without making organic chemistry impossible. Changing the quark masses will inevitably affect which baryons and which atomic nuclei can exist without decaying quickly. In turn, the different assortment of atomic nuclei will affect chemistry.

Quarky Chemistry

It seems plausible that intelligent life (if it is not very different from us) requires some form of organic chemistry, which is by definition the chemistry that involves carbon. The chemical properties of carbon follow from the fact that its nucleus has an electric charge of 6, so that six electrons orbit in a neutral carbon atom. These properties allow carbon to form an immense variety of complex molecules. (The suggestion often made by science-fiction writers that life could instead be based on silicon—the next element in carbon's group in the periodic table—is questionable: no silicon-based molecules of any significant degree of complexity are known to exist.) Furthermore, for complex organic molecules to form, elements with the chemistry of hydrogen (charge 1) and oxygen (charge 8) need to be present. To see if they could maintain organic chemistry, then, the team had to calculate whether nuclei of charge 1, 6 or 8 would decay radioactively before they could participate in chemical reactions.

The stability of a nucleus partly depends on its mass, which in turn depends on the masses of the baryons it is made of. Computing the masses of baryons and nuclei starting from the masses of the quarks is extremely challenging even in our universe. But after tweaking the intensity of the interaction between quarks, one can use the baryon masses measured in our universe to estimate how small changes to the masses of the quarks would affect the masses of nuclei.

In our world, the neutron is roughly 0.1 percent heavier than the proton. If the masses of the quarks were changed so that the neutron became 2 percent heavier than the proton, no long-lived form of carbon or oxygen would exist. If quark masses were adjusted to make the proton heavier than the neutron, then the proton in a hydrogen nucleus would capture the surrounding electron and turn into a neutron, so that hydrogen atoms could not exist for very long. But deuterium or tritium (hydrogen 3) might still be stable, and so would some forms of oxygen and carbon. Indeed, we found that only if the proton became heavier than the neutron by more than about 1 percent would there cease to be some stable form of hydrogen.

With deuterium (or tritium) substituting for hydrogen 1, oceans would be made of heavy water, which has subtly different physical and chemical properties from ordinary water. Still, there does not appear to be a fundamental obstacle in these worlds to some form of organic life evolving.

In our world, the third-lightest quark—the strange quark—is too heavy to participate in nuclear physics. But if its mass were reduced by a factor of more than about 10, nuclei could be made not just of protons and neutrons but also of other baryons containing strange quarks.

For example, the team identified a universe in which the up and strange quark would have roughly the same mass, whereas the down quark would be much lighter. Then atomic nuclei would not be made of protons and neutrons but instead of neutrons and another baryon, called the $\Sigma-$ ("sigma minus"). Remarkably, even such a radically different universe would have stable forms of hydrogen, carbon and oxygen and therefore could have organic chemistry. Whether those elements would be produced abundantly enough for life to evolve somewhere within them is an unanswered question.

But if life can arise, it would again happen much like it does in our world. Physicists in such a universe might be puzzled by the fact that the up and strange quarks would have almost identical masses. They might even imagine that this amazing coincidence has an anthropic explanation, based on the need for organic chemistry. We know, however, that such an explanation would be wrong, because our world has organic chemistry even though the masses of the down and strange quarks are quite different.

On the other hand, universes in which the three light quarks had roughly the same masses would probably have no organic chemistry: any nucleus with more than a couple of units of electrical charge would decay away almost immediately. Unfortunately, it is very difficult to map out in detail the histories of universes whose physical parameters are different from our own. This issue requires further research.

String Landscaping

Fine-tuning has been invoked by some theoretical physicists as indirect evidence for the multiverse. Do our findings therefore call the concept of a multiverse into question? We do not think that this is necessarily the case, for two reasons. The first comes from observation, combined with theory. Astronomical data strongly support the hypothesis that our universe started out as a tiny patch of spacetime, perhaps as small as a billionth the size of a proton, which then went through a phase of rapid, exponential growth, called inflation. Cosmology still lacks a definitive theoretical model for inflation, but theory suggests that different patches could inflate at different rates and that each patch could form a "pocket" that can become a universe in its own right, characterized by its own values for the constants of nature. Space between pocket universes would keep expanding so fast that it would be impossible to travel or send messages from one pocket to the next, even at the speed of light.

The second reason to suspect the existence of the multiverse is that one quantity still seems to be finely tuned to an extraordinary degree: the cosmological constant, which represents the amount of energy embodied in empty space. Quantum physics predicts that even otherwise empty space must contain energy. Einstein's general theory of relativity requires that all forms of energy exert gravity. If this energy is positive, it causes spacetime to expand at an exponentially accelerating rate. If it is negative, the universe would recollapse in a "big crunch." Quantum theory seems to imply that the cosmological constant should be so large—in the positive or negative direction—that space would expand too quickly for structures such as galaxies to have a chance to form or else that the universe would exist for a fraction of a second before recollapsing.

One way to explain why our universe avoided such disasters could be that some other term in the equations canceled out the effects of the cosmological constant. The trouble is that this term would have to be fine-tuned with exquisite precision. A deviation in even the 100th decimal place would lead to a universe without any significant structure.

In 1987 Steven Weinberg, the Nobel Prize–winning theorist at the University of Texas at Austin, proposed an anthropic explanation. He calculated an upper bound on the value of the cosmological constant that would still be compatible with life. Were the value any bigger, space would expand so quickly that the universe would lack the structures that life requires. In a way, then, our very existence predicts the low value of the constant.

Then, in the late 1990s, astronomers discovered that the universe is indeed expanding at an accelerating rate, pushed by a mysterious form of "dark energy." The observed rate implied that the cosmological constant is positive and tiny—within the bounds of Weinberg's prediction—meaning that dark energy is very dilute.

Thus, the cosmological constant seems to be fine-tuned to an exceptional degree. Moreover, the methods our teams have applied to the weak nuclear force and to the masses of quarks seem to fail in this case, because it seems impossible to find congenial universes in which the cosmological constant is substantially larger than the

Section 4: Searching for Answers

value we observe. Within a multiverse, the vast majority of universes could have cosmological constants incompatible with the formation of any structure.

A real-world analogy—as opposed to an action-movie one—would be to send thousands of people trekking across a mountainous desert. The few who make it out alive might tell stories full of cliffhangers, encounters with poisonous snakes, and other brushes with death that would seem too close to be realistic.

Theoretical arguments rooted in string theory—a speculative extension of the Standard Model that attempts to describe all forces as the vibrations of microscopic strings—seem to confirm such a scenario. These arguments suggest that during inflation the cosmological constant and other parameters could have taken a virtually limitless range of different values, called the string theory landscape.

Our own work, however, does cast some doubt on the usefulness of anthropic reasoning, at least beyond the case of the cosmological constant. It also raises important questions. For example, if life really is possible in a weakless universe, then why does our own universe have a weak force at all? In fact, particle physicists consider the weak force in our universe to be, in a sense, not weak enough. Its observed value seems unnaturally strong within the Standard Model. (The leading explanation for this mystery requires the existence of new particles and forces that physicists hope to discover at the newly opened Large Hadron Collider at CERN near Geneva.)

As a consequence, many theorists expect that most universes would have weak interactions that are so feeble as to be effectively absent. The real challenge, then, may be to explain why we do *not* live in a weakless universe.

Eventually only a deeper knowledge of how universes are born can answer such questions. In particular, we may discover physical principles of a more fundamental level that imply that nature prefers certain sets of laws over others.

We may never find any direct evidence of the existence of other universes, and we certainly will never get to visit one. But we may need to learn more about them if we want to understand what is our true place in the multiverse—or whatever it is that is out there.

About the Author

Alejandro Jenkins, a native of Costa Rica, is in the High Energy Physics group at Florida State University. He has degrees from Harvard University and the California Institute of Technology, and he investigated alternative universes while at the Massachusetts Institute of Technology with Bob Jaffe and Itamar Kimchi. Gilad Perez is a theorist at the Weizmann Institute of Science in Rehovot, Israel, where he received his Ph.D. in 2003. While at Lawrence Berkeley National Lab, he explored the multiverse with Roni Harnik of Stanford University and Graham D. Kribs of the University of Oregon. He has also done stints at Stony Brook University, Boston University and Harvard.

Does a Multiverse Fermi Paradox Disprove the Multiverse?

By Caleb A. Scharf

Having just orbited our way through another summer solstice, it feels like time to let slip some more speculative ideas before the hot days of the northern hemisphere shorten too much again and rational thinking returns.

So, grasping a fruity alcoholic beverage in one hand, consider the following thought experiment.

The so-called "Fermi Paradox" has become familiar fodder for speculations on the nature of life in the universe, so I'm not going to repeat it in any great detail here. Instead, take a look at this nice description by Adam Frank, and remember that the basic premise is: if life in the universe is not incredibly rare, it should have already shown up on our proverbial doorstep. The fact that is *hasn't* is therefore interesting.

But the universe is such a paltry thing. Hordes of physicists are telling us that our reality, our cosmos, may not be the only one–rather that we exist inside a *multiverse*.

This could be structured in a variety of forms: from pocket universes produced by cosmic inflation, to quantum mechanical diversions and "many-worlds," to "branes" in higher dimensional M-theory, and so on. Furthermore, all of these variants may not really be variants at all, they could all be mushed together into one stupendous array of realities. So many realities, in fact, that anything that can happen will (and must) happen, and will happen an enormous (dare I say, infinite) number of times.

Other hordes of physicists (well, perhaps not hordes, but a significant and sober segment of the physicist species) roll their eyes and point out the whiff of ludicrousness in some of this talk. After all, they say, theories that can explain absolutely anything you throw at them by just saying "anything is possible," are not exactly

Is There More Than One Universe?

theories according to the true scientific method, because they can't be rationally falsified. Touché.

Now, before you harangue me on behalf of one or the other side of this argument, take another slurp of your fruity cocktail and consider the following.

Let's suppose that the most liberal of multiverse ideas are true. In this case, a modern-day Fermi might find themselves making precisely the same statement that was made back in the 1950s: "Where is everyone?"

Except this time the question is not about where the interstellar travellers are, or why galactic civilizations haven't been spotted. The new puzzle is "Where are all the pan-multiverse travellers and civilizations?"

Have another sip, and let's unpack that question a little. If reality is actually composed of a vast, vast number of realities, and if "anything" can, does, and *must* happen, and happen many, many, times, this presumably has to include the possibility of living things (whatever they're composed of) skipping between universes willy-nilly. After all, just because physics in our universe makes that look kind of tricky, it doesn't prevent the physics of a huge number of other universes from saying "sure, go right ahead!"

And there's the rub. Discounting our all-too-human capacity for self-delusion, there is absolutely no hard evidence that we are being, or ever have been, visited by stuff from other realities. (And really, if you do feel inclined to comment and tell me I'm wrong about that, save your breath, sorry).

So what's the answer? Why *isn't* this happening?

It could be that traveling between parts of the multiverse is impossible (except it shouldn't be impossible everywhere, almost by definition) or very, very difficult.

It could be that no entity who reaches a stage where they could hop between universes actually wants to (except, there has to be someone somewhere who does. Again, almost by definition).

Section 4: Searching for Answers

It could be that we're alone, the only form of life in any reality (except, yet again, almost by definition, a multiverse will contain other life).

It could also be, very simply, that there is no multiverse.

Go on, drink up, that's what I'm doing.

The views expressed are those of the author(s) and are not necessarily those of Scientific American.

About the Author

Caleb A. Scharf is director of astrobiology at Columbia University. He is author and co-author of more than 100 scientific research articles in astronomy and astrophysics. His work has been featured in publications such as New Scientist, Scientific American, Science News, Cosmos Magazine, Physics Today and National Geographic. For many years he wrote the Life, Unbounded blog for Scientific American.

Was Our Universe Created in a Laboratory?

By Avi Loeb

The biggest mystery concerning the history of our universe is what happened before the big bang. Where did our universe come from? Nearly a century ago, Albert Einstein searched for steady-state alternatives to the big bang model because a beginning in time was not philosophically satisfying in his mind.

Now there are a variety of conjectures in the scientific literature for our cosmic origins, including the ideas that our universe emerged from a vacuum fluctuation, or that it is cyclic with repeated periods of contraction and expansion, or that it was selected by the anthropic principle out of the string theory landscape of the multiverse—where, as the MIT cosmologist Alan Guth says "everything that can happen will happen ... an infinite number of times," or that it emerged out of the collapse of matter in the interior of a black hole.

A less explored possibility is that our universe was created in the laboratory of an advanced technological civilization. Since our universe has a flat geometry with a zero net energy, an advanced civilization could have developed a technology that created a baby universe out of nothing through quantum tunneling.

This possible origin story unifies the religious notion of a creator with the secular notion of quantum gravity. We do not possess a predictive theory that combines the two pillars of modern physics: quantum mechanics and gravity. But a more advanced civilization might have accomplished this feat and mastered the technology of creating baby universes. If that happened, then not only could it account for the origin of our universe but it would also suggest that a universe like our own—which in this picture hosts an advanced technological civilization that gives birth to a new flat universe—is like a biological system that maintains the longevity of its genetic material through multiple generations.

Section 4: Searching for Answers

If so, our universe was not selected for us to exist in it—as suggested by conventional anthropic reasoning—but rather, it was selected such that it would give rise to civilizations which are much more advanced than we are. Those "smarter kids on our cosmic block"—which are capable of developing the technology needed to produce baby universes—are the drivers of the cosmic Darwinian selection process, whereas we cannot enable, as of yet, the rebirth of the cosmic conditions that led to our existence. One way to put it is that our civilization is still cosmologically sterile since we cannot reproduce the world that made us.

With this perspective, the technological level of civilizations should not be gauged by how much power they tap, as suggested by the scale envisioned in 1964 by Nikolai Kardashev. Instead, it should be measured by the ability of a civilization to reproduce the astrophysical conditions that led to its existence.

As of now, we are a low-level technological civilization, graded *class C* on the cosmic scale, since we are unable to recreate even the habitable conditions on our planet for when the sun will die. Even worse, we may be labeled *class D* since we are carelessly destroying the natural habitat on Earth through climate change, driven by our technologies. A *class B* civilization could adjust the conditions in its immediate environment to be independent of its host star. A civilization ranked *class A* could recreate the cosmic conditions that gave rise to its existence, namely produce a baby universe in a laboratory.

Achieving the distinction of class A civilization is nontrivial by the measures of physics as we know it. The related challenges, such as producing a large enough density of dark energy within a small region, already have been discussed in the scientific literature.

Since a self-replicating universe only needs to possess a single class A civilization, and having many more is much less likely, the most common universe would be the one that just barely makes class A civilizations. Anything better than this minimum requirement is much less likely to occur because it requires additional rare circumstances and does not provide a greater evolutionary benefit for the Darwinian selection process of baby universes.

Is There More Than One Universe?

The possibility that our civilization is not a particularly smart one should not take us by surprise. When I tell students at Harvard University that half of them are below the median of their class, they get upset. The stubborn reality might well be that we are statistically at the center of the bell-shaped probability distribution of our class of intelligent life-forms in the cosmos, even when taking into account our celebrated discovery of the Higgs boson by the Large Hadron Collider.

We must allow ourselves to look humbly through new telescopes, as envisioned by the recently announced Galileo Project, and search for smarter kids on our cosmic block. Otherwise, our ego trip may not end well, similarly to the experience of the dinosaurs, which dominated Earth until an object from space tarnished their illusion.

About the Author

Avi Loeb is former chair (2011-2020) of the astronomy department at Harvard University, founding director of Harvard's Black Hole Initiative and director of the Institute for Theory and Computation at the Harvard-Smithsonian Center for Astrophysics. He also chairs the Board on Physics and Astronomy of the National Academies and the advisory board for the Breakthrough Starshot project, and is a member of President's Council of Advisors on Science and Technology. Loeb is the bestselling author of Extraterrestrial: The First Sign of Intelligent Life Beyond Earth *(Houghton Mifflin Harcourt).*

Multiverse Theories Are Bad for Science

By John Horgan

In 1990 I wrote a bit of fluff for *Scientific American* about whether our cosmos might be just one in an "infinitude," as several theories of physics implied. I titled my piece "Here a Universe, There a Universe . . ." and kept the tone light, because I didn't want readers to take these cosmic conjectures too seriously. After all, there was no way of proving, or disproving, the existence of other universes.

Today, physicists still lack evidence of other universes, or even good ideas for obtaining evidence. Many nonetheless insist our cosmos really is just a mote of dust in a vast "multiverse." One especially eloquent and passionate multiverse theorist is Sean Carroll. His faith in the multiverse stems from his faith in quantum mechanics, which he sees as our best account of reality.

In his book *Something Deeply Hidden*, Carroll asserts that quantum mechanics describes not just very small things but everything, including us. "As far as we currently know," he writes, "quantum mechanics isn't just an approximation to the truth; it is the truth." And however preposterous it might seem, a multiverse, Carroll argues, is an inescapable consequence of quantum mechanics.

To make his case, he takes us deep into the surreal quantum world. Our world! The basic quantum equation, called a wave function, shows a particle—an electron, say—inhabiting many possible positions, with different probabilities assigned to each one. Aim an instrument at the electron to determine where it is, and you'll find it in just one place. You might reasonably assume that the wave function is just a statistical approximation of the electron's behavior, which can't be more precise because electrons are tiny and our instruments crude. But you would be wrong, according to Carroll. The electron exists as a kind of probabilistic blur until you observe it, when it "collapses," in physics lingo, into a single position.

Is There More Than One Universe?

Physicists and philosophers have been arguing about this "measurement problem" for almost a century now. Various other explanations have been proposed, but most are either implausible, making human consciousness a necessary component of reality, or kludgy, requiring *ad hoc* tweaks of the wave function. The only solution that makes sense to Carroll—because it preserves quantum mechanics in its purest form—was proposed in 1957 by a Princeton graduate student, Hugh Everett III. He conjectured that the electron actually inhabits all the positions allowed by the wave function, but in different universes.

This hypothesis, which came to be called the many-worlds theory, has been refined over the decades. It no longer entails acts of measurement, or consciousness (sorry New Agers). The universe supposedly splits, or branches, whenever one quantum particle jostles against another, making their wave functions collapse. This process, called "decoherence," happens all the time, everywhere. It is happening to you right now. And now. And now. Yes, zillions of your doppelgangers are out there at this very moment, probably having more fun than you. Asked why we don't feel ourselves splitting, Everett replied, "Do you feel the motion of the earth?"

Carroll addresses the problem of evidence, sort of. He says philosopher Karl Popper, who popularized the notion that scientific theories should be precise enough to be testable, or falsifiable, "had good things to say about" Everett's hypothesis, calling it "a completely objective discussion of quantum mechanics." (Popper, I must add, had doubts about natural selection, so his taste wasn't irreproachable.)

Carroll proposes furthermore that because quantum mechanics is falsifiable, the many-worlds hypothesis "is the most falsifiable theory ever invented"—even if we can never directly observe any of those many worlds. The term "many," by the way, is a gross understatement. The number of universes created since the big bang, Carroll estimates, is 2 to the power of 10 to the power of 112. Like I said, an infinitude.

Section 4: Searching for Answers

And that's just the many-worlds multiverse. Physicists have proposed even stranger multiverses, which science writer Tom Siegfried describes in his book *The Number of the Heavens*. String theory, which posits that all the forces of nature stem from stringy thingies wriggling in nine or more dimensions, implies that our cosmos is just a hillock in a sprawling "landscape" of universes, some with radically different laws and dimensions than ours. Chaotic inflation, a supercharged version of the big bang theory, suggests that our universe is a minuscule bubble in a boundless, frothy sea.

In addition to describing these and other multiverses, Siegfried provides a history of the idea of other worlds, which goes back to the ancient Greeks. (Is there anything they didn't think of first?) Acknowledging that "nobody can say for sure" whether other universes exist, Siegfried professes neutrality on their existence. But he goes on to construct an almost comically partisan defense of the multiverse, declaring that "it makes much more sense for a multiverse to exist than not."

Siegfried blames historical resistance to the concept of other worlds on Aristotle, who "argued with Vulcan-like assuredness" that earth is the only world. Because Aristotle was wrong about that, Siegfried seems to suggest, maybe modern multiverse skeptics are wrong too. After all, the known universe has expanded enormously since Aristotle's era. We learned only a century ago that the Milky Way is just one of many galaxies.

The logical next step, Siegfried contends, would be for us to discover that our entire cosmos is one of many. Rebutting skeptics who call multiverse theories "unscientific" because they are untestable, Siegfried retorts that the skeptics are unscientific, because they are "pre-supposing a definition of science that rules out multiverses to begin with." He calls skeptics "deniers"—a term usually linked to doubts about real things, like vaccines, climate change and the Holocaust.

I am not a multiverse denier, any more than I am a God denier. Science cannot resolve the existence of either God or the multiverse, making agnosticism the only sensible position. I see some value in

Is There More Than One Universe?

multiverse theories. Particularly when presented by a writer as gifted as Sean Carroll, they goad our imaginations and give us intimations of infinity. They make us feel really, really small—in a good way.

But I'm less entertained by multiverse theories than I once was, for a couple of reasons. First, science is in a slump, for reasons both internal and external. Science is ill-served when prominent thinkers tout ideas that can never be tested and hence are, sorry, unscientific. Moreover, at a time when our world, the real world, faces serious problems, dwelling on multiverses strikes me as escapism—akin to billionaires fantasizing about colonizing Mars. Shouldn't scientists do something more productive with their time?

Maybe in another universe Carroll and Siegfried have convinced me to take multiverses seriously, but I doubt it.

The views expressed are those of the author(s) and are not necessarily those of Scientific American.

About the Author

John Horgan directs the Center for Science Writings at the Stevens Institute of Technology. His books include The End of Science, The End of War *and* Mind-Body Problems, *available for free at mindbodyproblems.com. For many years he wrote the popular blog Cross Check for* Scientific American.

Life Quest: Could Parallel Universes Be Congenial to Life?

By Mariette DiChristina

After more than 40 years that included five long-running TV series (even an animated version) and a string of movies, the writers of the latest *Star Trek* blockbuster in theaters decided to move to a new universe—one that has created fresh opportunities for stories and the chance to modernize and update the franchise. In the movie last summer Kirk, Spock and the rest of the gang were back. But a critical change—a time-jumping, revenge-seeking madman who caused the death of Kirk's father and then destroyed the planet Vulcan—shattered the well-trod timeline of events that longtime fans have come to know so well.

Many *Star Trek* fans, old and new, like the new, parallel universe, which is intriguingly darker and gives beloved characters and the too-good-to-be-interesting Starfleet a helpful kick-start for future movies. One thing that struck me, however, was how similar the two universes actually were, aside from the cataclysms that brought forth the new timeline. They had the same starring roles (albeit with new, younger actors) and revolved around the same key worlds, the same Federation of Planets, and so on.

In science, as opposed to science fiction, parallel universes aren't necessarily so parallel. Beyond simple changes in character development, alternative universes may have wholly different laws of physics. Nevertheless, a number of them could prove to be congenial to life, which so far seems to be so rare in our own reality. According to prevailing cosmological theory, our universe spawned from a microscopic region of a primordial vacuum in a burst of exponential expansion called inflation; the vacuum may produce other universes as well. In numerous other universes, theorists long held, the laws of physics may not permit the formation of matter or galaxies as we know them—leaving our home unique.

Is There More Than One Universe?

But recent studies by Alejandro Jenkins and Gilad Perez, authors of our cover story, "Looking for Life in the Multiverse," show that some other universes may not be so inhospitable after all. "We have found examples of alternative values of the fundamental constants, and thus of alternative sets of physical laws, that might still lead to very interesting words and perhaps to life," they write. In other words, scientists get a "disaster" for life if their models vary just one "constant" of nature, but if they vary more than one they can find values that are compatible with the formation of complex structures and perhaps intelligent life. What would these universes be like?

Many of us are captivated by the search for other beings in the vast cosmos beyond Earth. So it is ironic that we sometimes place such a paltry value on life that already exists on our own planet. Seven horrific tropical diseases, mostly caused by parasitic worms, ruin the lives and health of a billion impoverished people around the world by making them chronically sick, yet these ailments get less attention and money than HIV/AIDS, malaria and tuberculosis. In his feature article, Peter Jay Hotez presents "A Plan to Defeat Neglected Tropical Diseases." Surely there is a way to provide the necessary drugs—which can cost just 50 cents per person—so that all people can thrive.

About the Author

Mariette DiChristina, Steering Group chair, is dean and professor of the practice in journalism at the Boston University College of Communication. She was formerly editor in chief of Scientific American *and executive vice president, Magazines, for Springer Nature.*

Multiverse Controversy Heats Up over Gravitational Waves

By Clara Moskowitz

The multiverse is one of the most divisive topics in physics, and it just became more so. The major announcement last week of evidence for primordial ripples in spacetime has bolstered a cosmological theory called inflation, and with it, some say, the idea that our universe is one of many universes floating like bubbles in a glass of champagne. Critics of the multiverse hypothesis claim that the idea is untestable—barely even science. But with evidence for inflation theory building up, the multiverse debate is coming to a head.

The big news last week came from the Background Imaging of Cosmic Extragalactic Polarization 2 (BICEP2) experiment at the South Pole, which saw imprints in the cosmic microwave background—the oldest light in the universe, dating from shortly after the big bang—that appear to have been caused by gravitational waves rippling through the fabric of spacetime in the early universe. The finding was heralded as a huge breakthrough, although physicists say confirmation from other experiments will be needed to corroborate the results.

If verified, these gravitational waves would be direct evidence for the theory of inflation, which suggests the universe expanded exponentially in the first fraction of a nanosecond after it was born. If inflation occurred, it would explain many features of our universe, such as the fact that it appears to be fairly smooth, with matter spread evenly in all directions (early inflation would have stretched out any irregularities in the universe).

Inflation might also mean that what we consider the universe—the expanse of everything we could see with the most perfect telescopes—is just one small corner of space, a pocket where inflation stopped and allowed matter to condense, galaxies and stars to form, and life

Is There More Than One Universe?

to evolve. Elsewhere, beyond the observable universe, spacetime may still be inflating, with other "bubble" universes forming whenever inflation stops in one location.

This picture is called eternal inflation. "Most inflationary models, almost all, predict that inflation should become eternal," says Alan Guth, a theoretical physicist at the Massachusetts Institute of Technology (MIT), who first predicted inflation in 1980.

If the BICEP2 results end up proving inflation occurred, then the multiverse may be part of the bargain. "I think the multiverse is a natural consequence of inflation ideas," says theoretical physicist Frank Wilczek, also at MIT. "If you can start one universe form a very small seed, then other universes could also grow from small seeds. There doesn't seem to be anything unique about the event we call the big bang. It is a reproducible event that could and would happen again, and again, and again."

If that is true, it could help explain why our universe seems so special. The mass of the electron, for example, appears to be completely random—this value is not predicted by any known physics. And yet if the electron were any heavier or lighter than it is, atoms could not form, galaxies would be impossible, and life would not exist. The same goes for many other constants of nature, especially the cosmological constant—the theorized, but unverified, source of the so-called dark energy that is propelling the acceleration of the expansion of the universe. If the cosmological constant were different, and dark energy was more or less powerful, the universe would be drastically altered, and life as we know it wouldn't be possible.

If our universe is the only one in existence, then we need some explanation for why it seems so fine-tuned for us to exist. If it is but one of many, however, then maybe each has different parameters, different constants, and one universe just happened to arrive at the values that enabled life.

"We live in this part of the universe because we can live there, not because the whole universe is built for our benefit," says Stanford University physicist Andrei Linde, one of the main authors of

Section 4: Searching for Answers

inflation theory and the multiverse hypothesis. This idea, called the anthropic principle, is satisfying to some, and maddening to others.

"That story gives a very neat and self-consistent picture," Guth says. But many find the anthropic principle and the multiverse distasteful. "The multiverse functions here as an all-purpose excuse for not being able to explain anything about particle physics," mathematician Peter Woit at Columbia University wrote in a blog post responding to BICEP2 reactions. "I consider such a view to be 'giving up' on finding a true scientific explanation," says Princeton University theoretical physicist Paul Steinhardt.

In addition to seeing the multiverse idea—and the anthropic principle it enables—as a cop-out, skeptics charge that it is impossible to test, because theory predicts other bubble universes would be permanently out of reach and unobservable. "Literally, anything can happen and does happen infinitely many times," Steinhardt says. "This makes the theory totally unpredictive or, equivalently, unfalsifiable."

An untestable idea is by definition unscientific, because science relies on verifying predictions through experimentation. Proponents of the multiverse idea, however, say it is so inextricable with some theories, including inflation, that evidence for one is evidence for the other.

"Once we have experimental proof that the cosmological constant is real, and we have experimental proof of inflationary cosmology, then suddenly we have something which I firmly believe is experimental evidence in favor of the multiverse," Linde says. "Those people who say the theory of the multiverse does not have any experimental confirmation have not paid enough attention."

Whether the BICEP2 results represent a piece of such confirmation is a point of contention between the pro- and anti-multiverse factions. "The BICEP2 discovery should cause dismay among multiverse skeptics—at least in this particular universe," MIT physicist Max Tegmark wrote in a guest blog for *Scientific American*.

Doubters, of course, vehemently disagree. "Perhaps there is a part of the multiverse in which the #BICEP2 results provide evidence for a multiverse, but I don't think we live there," Peter Coles, a theoretical cosmologist at the University of Sussex in England, wrote on Twitter.

And many physicists are agnostic about what, if anything, the BICEP2 results have to say about the multiverse. "The multiverse is an idea for how the inflationary period that gave rise to our universe may have come about," says Marc Kamionkowski, professor of physics and astronomy at Johns Hopkins University. "This particular measurement doesn't shed any direct light on that."

Ultimately, neither side of the debate is likely to concede defeat any time soon. But one faction at least is claiming a small victory from last week's news. "The more we move in this way, the more seriously we should take the possibility of eternal inflation and the multiverse," Linde says, "and the idea that our universe is not just one cosmic balloon but a fractal of balloons producing new balloons producing new balloons forever."

About the Author

Clara Moskowitz is Scientific American's *senior editor covering space and physics. She has a bachelor's degree in astronomy and physics from Wesleyan University and a graduate degree in science journalism from the University of California, Santa Cruz.*

GLOSSARY

analogous Similar in some way.

cosmology The scientific study of the origin and structure of the universe.

dark energy A hypothetical form of energy that produces a force that opposes gravity.

empirical Based on testing or experience.

entropy The breakdown of the matter and energy in the universe to an ultimate state of inert uniformity.

eponymous Relating to the person or thing for whom something is named.

event horizon The boundary of a black hole beyond which nothing can escape from within it.

fallacy An often plausible argument using false or invalid inference.

infinitesimal Extremely small.

inflation A hypothetical brief period of very rapid expansion of the universe immediately following the Big Bang.

multiplicity A very large number of a thing.

paradox Something that is made up of two opposite things and that seems impossible but is actually true or possible.

parameter A rule or limit that controls what something is or how something should be done.

pedagogical Of or relating to teachers or education.

Planck's constant The mathematical formula which describes the behavior of the smallest pieces of matter known to exist.

quantum entanglement The phenomenon that occurs when a group of particles are generated, interact, or share proximity.

quantum mechanics A branch of physics that deals with the structure and behavior of very small pieces of matter.

relativity A theory which says that the way that anything except light moves through time and space depends on the position and movement of someone who is watching.

superposition The combination of two phenomena of the same type so that they coexist as part of the same event.

FURTHER INFORMATION

Ananthaswamy, Anil. "Is Our Universe a Hologram? Physicists Debate Famous Idea on Its 25th Anniversary," *Scientific American*, March 1, 2023, https://www.scientificamerican.com/article/is-our-universe-a-hologram-physicists-debate-famous-idea-on-its-25th-anniversary1/.

Falk, Dan. "Scientists Attempt to Map the Multiverse," *Discover*, February 10, 2023, https://www.discovermagazine.com/the-sciences/scientists-attempt-to-map-the-multiverse.

Huang, Nancy. "Parallel Universe Theory: What Are the Chances of Another You?" *Now*, February 24, 2023, https://now.northropgrumman.com/parallel-universe-theory-what-are-the-chances-of-another-you/.

Kaku, Michio. "In a Parallel Universe, Another You," *The New York Times*, June 20, 2022, https://www.nytimes.com/2022/06/20/special-series/michio-kaku-multiverse-reality.html.

Saikrishna, Aditya. "String Theory: The Theory That Ties Everything Together," *Transcontinental Times*, February 16, 2023, https://www.transcontinentaltimes.com/string-theory-ties-everything/.

Siegel, Ethan. "Is There Another 'You' Out There In A Parallel Universe?" *Forbes*, November 18, 2016, https://www.forbes.com/sites/startswithabang/2016/11/18/is-there-another-you-out-there-in-a-parallel-universe/?sh=1465e37a634f.

Stein, Vicky and Daisy Dobrijevic. "Do parallel universes exist? We Might Live in a Multiverse," *Space.com*, November 3, 2021, https://www.space.com/32728-parallel-universes.html.

Wells, Sarah. "A Quantum Explanation for Gravity Could Generate the Theory of Everything," *Popular Mechanics*, February 10, 2023, https://www.popularmechanics.com/science/a42736230/is-quantum-gravity-real/.

M

Merali, Zeeya, 37–41, 74–83
Michalakis, Spyridon, 43–58
Moskowitz, Clara, 122–126, 151–154
multiverse, 8–14, 16, 23, 26–33, 84–93, 122–124, 128–142, 145–148, 150–154
Musser, George, 34–36

N

Nomura, Yasunori, 84–93

O

observables, 62, 66, 69

P

Perez, Gilad, 128–138, 150

Q

quantum physics/mechanics, 11–12, 14, 35, 37–41, 43–66, 68–77, 80–83, 85–98, 100–101, 103–105–113, 115, 118–120, 123, 136, 139, 142, 145–146
Quevedo, Fernando, 16–25

R

relativity, 11, 59–61, 63–64, 68, 71, 78, 98, 108, 115, 123, 136

S

scalar field, 18, 21
Scharf, Caleb A., 139–141
Shaghoulian, Edgar, 98–106
Siegfried, Tom, 26–29, 147–148
spacetime, 19, 24, 28, 61–62, 88–90, 99, 104, 107–108, 112–115, 122–124, 135, 151
string theory, 8–9, 11–12, 16–17, 19–21, 23–25, 33, 68, 104–105, 118–126, 137, 142, 147
superposition, 70, 74–79, 81–82, 86–87, 90–91, 94–95, 109–110, 113
supersymmetry, 34–35

T

Tegmark, Max, 10–15, 34, 153
time, 19, 24, 37–41, 43, 60–62, 66–69, 85, 92, 107, 109
topology, 37–39, 41, 49–53, 57, 64–65, 67, 69

V

Vilenkin, Alexander, 7–9

W

wave function, 70–71, 145–146
wormhole, 62, 67, 112–115